2-6-64

RICE UNIVERSITY

SEMICENTENNIAL PUBLICATIONS

The Earth Sciences

EDITOR

THOMAS W. DONNELLY

CONTRIBUTORS

SYDNEY P. CLARK, JR.

JOHN A. O'KEEFE

W. S. FYFE

FRED A. DONATH

S. S. WILKS

HEINZ A. LOWENSTAM

The Earth Sciences

PROBLEMS AND PROGRESS IN CURRENT RESEARCH

PUBLISHED FOR
WILLIAM MARSH RICE UNIVERSITY
BY
THE UNIVERSITY OF CHICAGO PRESS

Library of Congress Catalog Card Number: 63-20901

THE UNIVERSITY OF CHICAGO PRESS, CHICAGO & LONDON
The University of Toronto Press, Toronto 5, Canada

Published 1963. Composed and printed by THE UNIVERSITY OF CHICAGO PRESS, *Chicago, Illinois, U.S.A.*

Contents

Introduction

During the academic year 1962–63 Rice University has been celebrating its fiftieth birthday. Most of the departments of the university have participated in this celebration by inviting distinguished scholars to address the university and to prepare for publication essays in their fields of interest. As part of its participation in the semicentennial celebration, the Department of Geology invited six scientists to prepare lectures on what they considered to be important and exciting fields of current research in the earth sciences. These were initially delivered during a symposium entitled "Frontiers of Geology," held at the university on November 15, 1962, in conjunction with the annual meeting in Houston of the Geological Society of America.

Because these lectures were delivered on the occasion of Rice's fiftieth birthday, it might be interesting to turn back the clock fifty years and speculate as to the contents of a hypothetical series of lectures that might have been held on the occasion of Rice's opening, if Rice had had a geology department at that time. J. F. Hayford might have been asked to lecture on the importance of isostasy—a concept that he had recently discussed at length in several long papers. Someone might have suggested that the young physicist Percy Bridgman prepare a lecture on his work with high pressures. One of the scientists from the newly founded Geophysical Laboratory—perhaps the young Canadian geologist N. L. Bowen—might have been asked to present a talk on the concept of phase equilibrium studies and their bearing on petrological problems. And many people would have been interested in hearing what F. D. Adams had to say about experiments with rock deformation and their relation to rock mechanics and structural geology. If the organizing committee had had sufficient funds, it might have invited John Murray, whose book on oceanography had just been published, to discuss the present state and future of this science. Possibly someone with sufficient foresight might have invited Max von Laue to speculate on the impact of X-ray diffraction on the study of crystals, though the significance of this field of study was not apparent to most scientists until a few years later. These examples are not completely representative of scientific interest prior to World War I, but it is not unfair to say that each of these topics was of great interest to a large number of scientists at that time. More significant is the probability that few geologists would have been familiar with either von Laue's or Bridgman's work then, though both were to prove to be of fundamental importance to earth science later.

The foregoing list of hypothetical topics was chosen for a particular purpose—to demonstrate that many of the problems discussed by the present group of speakers were not really new problems in 1962. They were not really new in 1912, either. As early as 1859 Sorby emphasized the importance of using thin sections to discern the original organic structures of fossils. In 1864 Thomson refuted some earlier notions as to the earth's heat and speculated as to the time of accretion of the earth, which he believed to have been a few hundred million years ago. In 1878 Daubrée attempted to elucidate structural processes by laboratory experiments—finding, among other things, that thickness of strata plays a very important role in folding. The most important problems, as W. S. Fyfe states, are also the oldest. We are not asking completely new questions today, but we are asking some of the old questions in a newer and, we hope, more profitable way.

If current research does not revolve around completely new questions, then we might ask in what way it does differ from the research of 1912. One view is that today's research is quantitative, and that of 1912 qualitative. This statement is only partially true. Certainly, sophisticated analyses of numerical data according to what we now recognize as statistical techniques were notably lacking in the earth sciences of fifty years ago, but there is no paucity of numerical data in the literature of that time. There is no indication that the pre–World War I geologist was hesitant about measuring quantities of all sorts and recording those measurements. The evolution of the earth sciences during the last fifty years can better be described as an ever increasing integration with the other sciences. Today's geologist is not the provincial scientist that his predecessor was but is more cognizant of the concepts of chemistry, physics, mathematics, biology, and even astronomy—and is more aware of their bearing on the earth sciences. It is highly likely that the significance of the work of today's von Laues and Bridgmans will not remain unnoticed for long by earth scientists. The reader might notice the backgrounds of the contributors to this volume—only two had "classical" geological educations—or the percentage of the book devoted to problems of other sciences, and he might inquire as to where the geology is. The answer is simple: This is today's geology!

The entire staff of the Department of Geology has participated in the planning of this symposium. Professors John J. W. Rogers and B. C. Burchfiel have assisted with the critical editing of the manuscripts. Finally, an especial debt of gratitude is owed Chancellor Carey Croneis, chairman of the department and of the university's semicentennial committee, for enabling this symposium to be presented.

SYDNEY P. CLARK, JR.

Variation of Density in the Earth and the Melting Curve in the Mantle

RATHER THAN ATTEMPT to cover the whole broad subject of the nature of the earth's interior, it seems preferable to restrict this review to two special topics in which considerable progress has been made in recent years. These are the earth's density variation and the melting curve in the mantle. Other topics important to the theory of the interior of the earth are the results of analysis of the orbits of artificial satellites and the rheological properties of rocks; these are considered in other chapters of this volume.

The principal reason for re-examining the density variation arises from recent laboratory work at high pressures and temperatures. A strong correlation between density and the velocity of compressional elastic waves has been demonstrated in materials of similar mean atomic weights. This implies that regions in the earth in which the density rises rapidly with increasing depth should be regions in which the velocity behaves in a similar fashion. This places additional limits on the density distribution.

A second important advance in the laboratory has been the semiquantitative prediction of the existence and properties of high-pressure phases in common silicate systems and, in some cases, their synthesis. This work has changed to a hypothesis that is correct beyond reasonable doubt the originally rather vague suggestion that the properties of the outer mantle are closely related to changes of phase. Of special importance is the discovery of stishovite, a new, dense polymorph of silica with the rutile structure. A mixture of oxides, with silica present as

Dr. Clark was born in Philadelphia and attended Harvard University, where he received the Bachelor's and Doctor's degrees, the latter in 1955. Following two years' postdoctoral work in the Dunbar Laboratory at Harvard, he joined the staff of the Geophysical Laboratory in Washington, D.C. He has recently been appointed the Sidney Weinberg Professor of Geophysics at Yale University. His principal interest has been the physics of the earth's interior.

stishovite, is denser than any of the common rock-forming silicates; hence decompositions to oxides represent a possible series of phase transformations in the mantle. The volume changes associated with these and other inversions to high-pressure phases can be estimated crudely and can be compared with the anomalous rise in density in the upper mantle indicated by geophysical results. In this way the question of whether the mantle is chemically, as well as physically, inhomogeneous can be approached.

The recognition that phase changes are important in the outer part of the mantle implies that a re-examination of the melting curve in the mantle is needed. When a solid-solid transition intersects the melting curve in the pressure-temperature plane, the melting curve is refracted, and its slope, dT/dP, increases with increasing pressure. This is opposite to the normal tendency of the slopes of melting curves to decrease with increasing pressure, and it implies that the gradient of the melting point in the lower part of the mantle is greater than might be supposed in the absence of phase transitions.

The basic data upon which conclusions about the interior of the earth are drawn are still in a crude, preliminary state in many cases. The advance in knowledge that has taken place over the past decade is certainly impressive, but nonetheless, commonly, it has simply meant that results that may be regarded as roughly known within broad limits are now available in lieu of no data at all. Approximations that are known not to be strictly correct must still be used if one is to reach a solution, and the resulting uncertainty is virtually impossible to assess. One such simplification, which pervades all the work discussed below, is the treatment of the earth as a spherically symmetrical body. This is certainly not true, but geophysical observations have generally not, up to the present, been sufficiently extensive or accurate to reveal the dependence of the various properties of the earth on latitude and longitude. An attack on this problem has begun, and it is likely to be one of the major geophysical advances of the present decade. The explosive growth in the field of data-processing is a major reason for its feasibility.

Seismic and Geodetic Data

The fundamental quantities pertaining to the density distribution are the P and S velocities in the interior and the earth's mass and moment of inertia. As the seismic velocity distributions became reasonably well known, seismologists found that they could recognize six main subdivisions of the mantle and core. These were based mainly on Jeffreys' (1939*a*, 1939*b*) solutions for the velocities. The subdivisions are described by Bullen (1947; see also Birch, 1952). Later work, notably that by Gutenberg, indicated that the boundary between two of the layers in the upper mantle was characterized by no seismic discontinuity of any sort. Hence, there is little basis for their distinction. Furthermore, one of the layers formerly recognized in the core is probably nonexistent. Gutenberg's last results (published by Bullard, 1957) have been confirmed in the upper mantle by studies of surface waves (Dorman, Ewing, and Oliver, 1960) and by analysis of the free oscillations of the earth (MacDonald and Ness, 1961). Gutenberg's velocities are shown in figure 1.

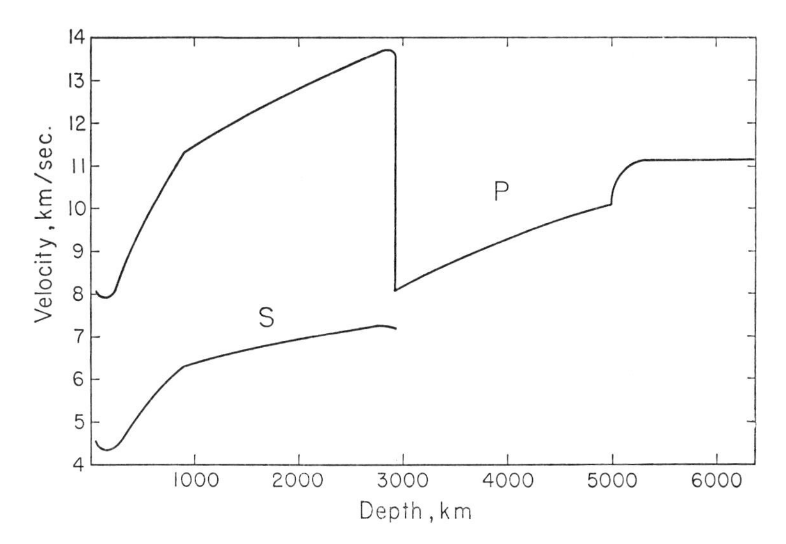

FIG. 1.—Gutenberg's latest velocities as given by Bullard (1957)

The modern view of the earth, which is adopted here, recognizes a twofold subdivision of both the mantle and the core. The upper mantle extends from the Mohorovičić discontinuity to a depth of approximately 1,000 km. It is characterized by velocities decreasing with increasing depth near its outer margin, passing through a minimum at a depth of about 160 km., and finally rising sharply (fig. 1). Gutenberg's final solutions show a second-order discontinuity (break in slope of the velocity-depth curve) at a depth of 900 km. It is not clear that such a feature is demanded by the travel times; it seems equally probable that the velocity gradient decreases near the base of the upper mantle and that the velocity-depth curves are smooth in this region. The older Jeffreys velocities showed no discontinuities at depths near 1,000 km.

The velocities in the lower mantle increase gently and smoothly with increasing depth throughout most of this region. Within about 200 km. of the core boundary, however, the velocity gradients decrease to zero or even change sign in both Jeffreys' and Gutenberg's solutions. The significance of this feature, if real, remains in doubt.

In the outer core the P velocity rises gently and smoothly. The margin of the inner core is marked by an abrupt, but not discontinuous, increase in velocity. The P velocity in the inner core passes through a maximum and actually decreases toward the center, according to Gutenberg's solution. This seems unlikely; the uncertainties in the velocities in this small central region are very large. No S waves are observed in the core, and this is the principal evidence that the outer core, at least, is liquid.

The mass of the earth is determined from its radius, observed gravity at the surface, and the universal constant of gravitation, G. Of these quantities, G is the least precisely known; its uncertainty is slightly less than 1 part in 1,000. The mass of the earth is found to be 5.977×10^{27} gm., with an uncertainty nearly the same as that in G.

The earth's moment of inertia is determined from the precessional constant and the ellipticity of the surface. The former is well determined from observations of the precession of the equinoxes. For many years the best value of the flattening coefficient f, which equals $(a_e - a_p)/a_e$ (a_e and a_p are the equatorial and polar radii, respectively), came from adjustment of geodetic arcs and measurements of gravity. These led to a value of f^{-1} close to 297; Jeffreys' latest value was 297.10 ± 0.36 (Jeffreys, 1959, p. 193). The stated uncertainty is the standard error, but Jeffreys (1959, p. 154) believed that further geodetic observations might require a change in this value by as much as 1 part in 300. It is now possible to calculate f from the influence of the equatorial bulge on the orbits of artificial satellites. Observations leading to the determination of the orbital elements started in 1957. The most complete sets of observations and reductions are available for the satellites 1957β (Sputnik II) and 1958β (Vanguard I); they lead to $f^{-1} = 298.24 \pm 0.01$ (Kaula, 1961).[1] Orbits of other satellites, which have been less thoroughly studied, lead to values of f that differ from this by a few parts in 10,000 (King-Hele, 1961). Further accumulation of observations will apparently be necessary before the precision of the satellite value of f can be finally decided. It appears, however, that the f as deduced from the satellite data is distinctly smaller than the geodetic value. It leads to a moment of inertia of 8.019×10^{44} gm. cm.^{2}. The value adopted by Bullen in a classic series of papers on the density variation was 8.104×10^{44} gm. cm.^{2}. The reduced moment of inertia implies that the mantle is, on the average, lighter, and the core heavier, than in Bullen's models.

[1] [EDITOR'S NOTE: Compare with the value given by O'Keefe in this volume.]

The Earth's Density Variation

The usual approach to the density variation is through the equations describing the self-compression of a homogeneous layer. If the thermal gradient is adiabatic, this leads to the Williamson-Adams relation, which is the most powerful means of deducing the density variation in regions where it is applicable. It is not valid in chemically or physically inhomogeneous regions or, in its original form, in regions in which the thermal gradient is not adiabatic. An allowance for departure from adiabaticity can easily be made, but it is approximate in the sense that one of the quantities involved, the thermal expansion, must be estimated.

This sort of investigation starts with the seismic velocities. By their definition in terms of the elastic constants of an isotropic medium

$$V_P^2 - 4/3\,V_S^2 = K_S/\rho \equiv (\partial P/\partial\rho)_S \equiv \phi\,, \tag{1}$$

where K_S is the adiabatic bulk modulus (the reciprocal of the adiabatic compressibility), P is pressure, and ρ is density.[2] On the assumption of hydrostatic pressures, which is valid to high precision deep in the earth, where the mean pressure greatly exceeds the shearing strength of the material, $dP/dr = -g\rho$, where r is radius and g is the gravitational acceleration, we have

$$d\rho = (\partial\rho/\partial P)_T dP + (\partial\rho/\partial T)_P dT \tag{2}$$

and

$$d\rho/dr = -\,g\rho^2/K_T - \rho\alpha\, dT/dr\,, \tag{3}$$

where K_T is the isothermal bulk modulus and α is the thermal expansion. $K_T/K_S = 1 - T\alpha^2 K_T/\rho C_P$, where C_P is the heat capacity at constant pressure. Now, in general, the thermal gradient can be split into the adiabatic gradient and a non-adiabatic component. The former equals $-T\alpha g/C_P$ and the latter is denoted by τ, where the sign of τ is such that it is positive when the increase of temperature with increasing depth exceeds the adiabatic. Equation (3) then becomes

$$d\rho/dr = -\,g\rho^2/K_T + T\alpha^2 g\rho/C_P + \alpha\rho\tau$$

or

$$d\rho/dr = -\,g\rho/\phi + \alpha\rho\tau\,. \tag{4}$$

Equation (4), with $\tau = 0$, is the Williamson-Adams relation, originally used by Williamson and Adams (1923) in an early study of the density variation. The generalization to non-adiabatic conditions is due to Birch (1939, 1952).

Given the seismic velocities, it is possible to integrate the William-

[2] Conventional notation leads to a slight ambiguity here. V_S refers to shear velocity, whereas the subscript S refers to adiabatic conditions in the other terms.

son-Adams equation, starting from, say, an assumed density at the top of the mantle. This is what Williamson and Adams set out to do, but they found that for any reasonable density at the top of the mantle the resulting earth model was much less massive than the actual earth. The discrepancy could be removed by assuming the presence of a dense metallic core. Using later, and greatly improved, velocities, Bullen (1936) repeated the integration to the core boundary, using 3.32 gm/cm^3 for the density at the top of the mantle. This led to a value of z $(=I/MR^2)$, of 0.57 for the core. (Here I denotes moment of inertia, M is mass, and R is radius; z is a dimensionless number.) The value of z is 0.4 for a uniform sphere and 0.67 for a spherical shell. Bullen's result implies that the density of the core rapidly decreases inward, which is unacceptable in view of the low rigidity of the core. This difficulty can be removed by adopting a higher density at the top of the mantle, but a value of 3.7 is required (Birch, 1953), and this is unacceptably high. Bullen escaped from the difficulty first by assuming a discontinuous jump in density at 400 km. (Bullen, 1936) and later by assuming that the density increased continuously at a rate exceeding that given by the Williamson-Adams relation between 400 and 1,000 km. (Bullen, 1940).

These results indicate inhomogeneity in the mantle. Birch (1952) examined the question in detail and reached the conclusion that the Williamson-Adams method was not valid in the upper mantle both because of high thermal gradients and because of physical, and perhaps chemical, inhomogeneity. Some other way of deducing the density in that region must be found. In the lower mantle, on the other hand, Birch's analysis indicated that the material was homogeneous as nearly as could be judged, except for the thin layer near the boundary of the core. The seismic results in that region could be attributed to an increasing content of iron with depth. Throughout the greater part of the lower mantle, which is the largest and most massive subdivision of the earth, equation (4) should apply. It is not certain whether τ is zero in the lower mantle.

In the outer core the material is presumably liquid, and hydromagnetic theories of the origin of the earth's magnetic field require that it be in motion (see, for example, Hide, 1956). It is implausible that chemical heterogeneity could persist in such a region, and convective motions would prevent the thermal gradient from exceeding the adiabatic. The Williamson-Adams relation should hold there.

Discussion of the homogeneity of the inner core is complicated by the uncertainties in the velocity distribution. The situation is illustrated in figure 2, taken from Gutenberg (1958). Only a small part of the path of inner-core waves is in the inner core at these large epicentral dis-

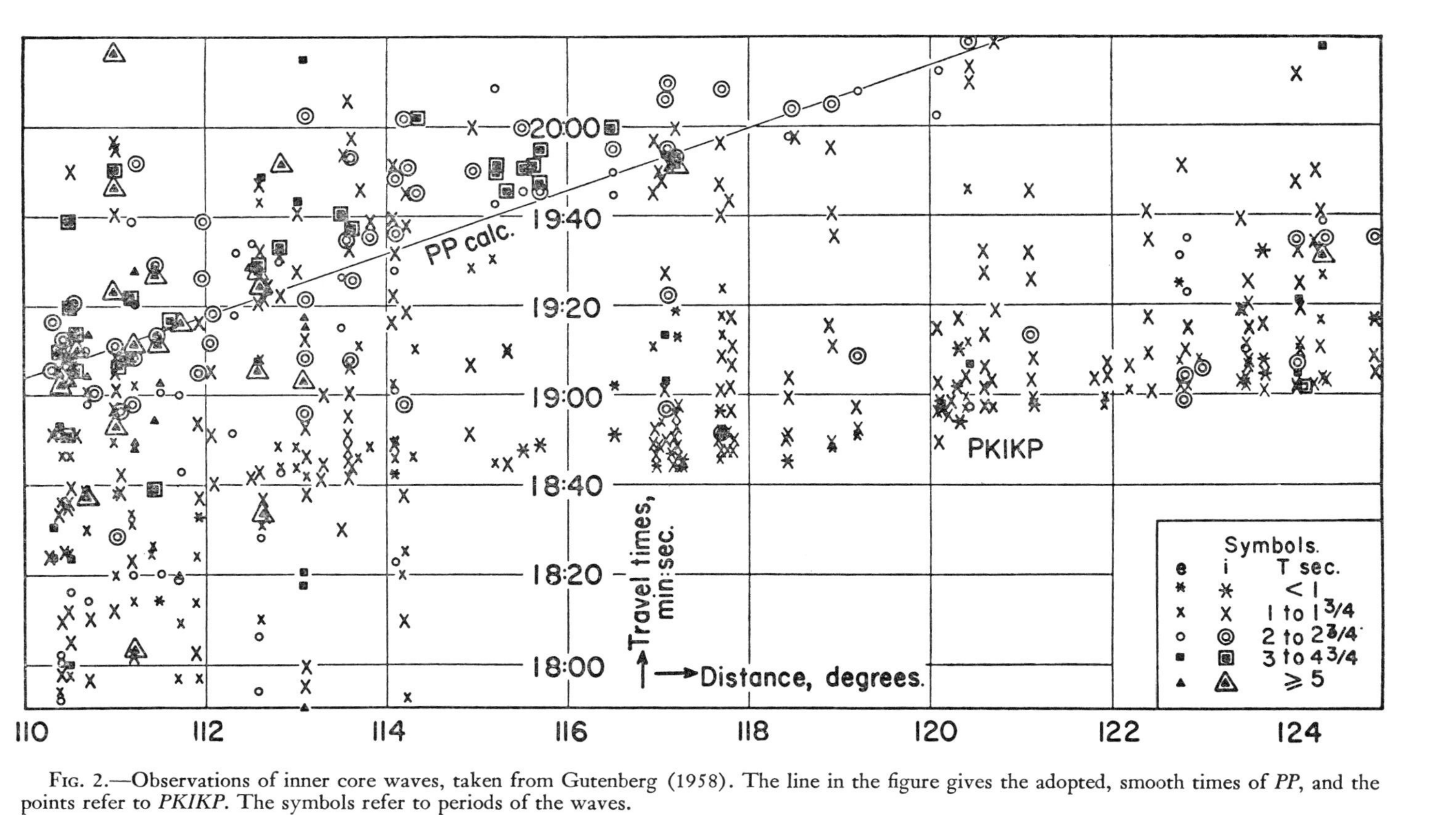

FIG. 2.—Observations of inner core waves, taken from Gutenberg (1958). The line in the figure gives the adopted, smooth times of *PP*, and the points refer to *PKIKP*. The symbols refer to periods of the waves.

tances, and the large scatter of the times shown in figure 2 may partially arise from regional differences in the upper mantle, for example. Much scatter is of course related to errors in the location of epicenters and in the origin times. Future work may lead to more accurate velocities in the inner core, but this will evidently be a matter of considerable difficulty in this small central region. But, because of its small size, the inner core comprises only roughly 2 per cent of the mass of the earth and less than 0.1 per cent of its moment of inertia. Hence, changing its properties produces very minor shifts in the properties of other layers.

We now turn to a detailed discussion of the density distribution. All the calculations described below were done by an IBM 7090 digital computer, using programs written in Fortran. It proves to be convenient to consider some of the properties of the core at the outset.

Two principal theories about the chemical composition of the earth's core have been seriously entertained in the past decade. The first is the classical view that it is dominantly nickel-iron, as suggested by the composition of iron meteorites. An alternative hypothesis, due to Ramsey (1948), is that the core is a high-pressure, metallic modification of the silicate material that makes up the mantle. It appears on the basis of later work that Ramsey's hypothesis must be discarded. Knopoff and MacDonald (1960) have concluded from Thomas-Fermi theory and shock-wave data that the mean atomic number of the core material is greater than 20. Birch (1961*a*) has compared the velocity-density relation in the core with those of a number of metals under shock compressions and has concluded that the core is most nearly comparable to metals of the first transition series. No reasonable adjustment of the density data can change this conclusion. Studies of the abundances of elements in meteorites and in the solar chromosphere (Aller, 1961) show that no element heavier than those in the first transition series has more than one-thousandth the abundance of iron. These data give no support to the notion that the mean atomic number of the core is maintained at a relatively high value through the addition of heavy elements to a material having the mean atomic number of ordinary silicates. Nor do they support the idea that the inner core is composed of material that is appreciably heavier than iron. There is evidence that silicon may be dissolved in the alloy comprising the outer core (Ringwood, 1959; MacDonald and Knopoff, 1958). It may also contain significant amounts of carbon and sulfur.

The simplest view of the inner core is that it consists of material similar to the outer core, but solidified by the higher pressure. One alternative hypothesis is that the inner core is also liquid, but denser than the outer core because of electronic collapse. The work of Newton, Jayaraman, and Kennedy (1962) on the alkali metals suggests that this process leads to broad, vague, transition zones in liquids, if this is indeed the correct interpretation of their results for cesium and Bundy's (1959) data for rubidium. It seems most unlikely that a transition of this sort could give rise to the velocity distribution found by Gutenberg, although in view of the various uncertainties, including the seismic ones, perhaps this alternative hypothesis should not be abandoned entirely. It does not seem as plausible as the idea that the inner core is solid.

Then, the solidity of the inner core being assumed, it is necessary to attempt to estimate the magnitude of the jump in density accompanying solidification of the core at the relevant physical conditions. Data for the chemically complex material that doubtless actually makes up the core are not available. Information about the properties of pure iron is available, and, since it appears to be the dominant material in the core, its properties will presumably provide a guide the to actual situation.

A solid iron-rich alloy at the pressure and temperature of the inner core presumably has the γ (face-centered) structure. This structure is stabilized by nickel and carbon, whereas silicon in solid solution increases the range of stability of the δ (body-centered) structure. It appears, however, that the effect of pressure would dominate that of composition, and pressure favors the denser, face-centered phase. The high-temperature X-ray diffraction data of Basinski *et al.* (1955), when combined with the thermodynamic data summarized by Kelley (1949), yield a slope of about 15 deg/kb for the γ-δ phase boundary. When Strong's (1959) data on the fusion curve of iron are converted to the currently accepted high-pressure scale, they determine an essentially straight line with a slope of 2.56 deg/kb. These data indicate a triple point at 10–12 kb., with γ-iron stable at the liquidus at higher pressures.

The heat of fusion of γ-iron at its metastable melting point can be obtained by using the heat contents and heat capacities of γ-iron and liquid iron given by Kelley (1949). The result is 18.500 j/mole at a melting point assumed to be 1,520° C. Analysis of the data of Basinski *et al.* (1955) shows that γ-iron has a volume of 7.549 cm^3/mole at 1,388° C. and a mean volumetric thermal expansion of 7.15×10^{-5}/deg. The volume of liquid iron is 7.98 cm^3/mole at 1,564° C., and the thermal expansion is variously estimated to be between 7 and 14×10^{-5}/deg (Stott and Rendall, 1953). With these data, and assuming a metastable melting point of 1,520° C., we find that the volume change on fusion lies between 0.31 and 0.34 cm^3/mole. With an entropy of fusion of 10.3 j/mole deg, derived from the thermodynamic data, the predicted slope of the melting curve lies between 3 and 3.3 deg/kb, a value that is not in particularly good agreement with Strong's result of 2.56 deg/kb. The discrepancy is no larger than is commonly found in this sort of calculation, however, and may in part arise from the fact that the thermodynamic data give the slope of the melting curve at zero pressure, whereas Strong measured the mean slope at high pressures. While Strong's data can be fitted with a straight line, they are not precise enough to rule out the possibility of slight curvature. The discrepancy would be reduced by taking a smaller volume change, and these results make it plausible that γ-iron melts with a volume change that is, at most, 5 per cent of the volume of the liquid.

An estimate of the effect of pressure on the volume change can be obtained through Simon's equation for the melting curve. This is

$$P = A\,[(T/T_0)^c - 1], \tag{5}$$

where T_0 is the melting point at zero pressure and A and c are disposable constants. Differentiation of equation (5) yields

$$\left(\frac{\Delta V}{\Delta S}\right)_P = \left(\frac{\Delta V}{\Delta S}\right)_0 (P/A + 1)^{1/c-1}, \tag{6}$$

where the subscripts 0 and P refer to the respective pressures, and ΔV and ΔS are, respectively, the changes of volume and entropy upon fusion.

Since c is greater than 1 in known cases, equation (6) predicts that ΔV decreases along the melting curve unless ΔS increases. Bridgman (1952, p. 201) states that in every case known to him ΔS decreases with pressure along the melting curve. Such theory as there is points in the same direction. The order-disorder theory of melting of Lennard-Jones and Devonshire (1939) actually predicts that the melting curve ends in a critical point, at which both ΔV and ΔS must vanish. These authors are careful to remark that no great significance should be attached to this result of an approximate theory, and it is not in good agreement with experiment. Nevertheless, the theory indicates that ΔS decreases with pressure. Bridgman (1952, p. 198, 203) was of the opinion that both ΔV and ΔS decreased along the fusion curve but that neither necessarily vanished at any finite pressure.

A further argument on this question emerges from the extensive discussion of melting by Slater (1939, p. 260). He separates the entropy change on fusion into a configurational term—approximated by assuming a random distribution of holes in an otherwise regular lattice—and a term expressing the dependence of the difference in entropy between solid and liquid on the difference in volume. This latter term is evaluated by integration of the specific heats of solid and liquid, calculated from the Debye model; it accounts for a far greater part of the entropy change than does the configurational term. Slater's theory should not be considered to have great quantitative significance, since it is admittedly only a first, very rough approximation, but it is a further indication that ΔS does indeed decrease with increasing pressure along the fusion curve, largely because of the decrease in ΔV.

The initial change in ΔS along the fusion curve of iron can be calculated from atmospheric pressure data on thermal expansion and heat capacity. The results suggest that ΔS does indeed decrease initially, but the small difference in heat capacity between solid and liquid and the large uncertainty in the thermal expansion of liquid iron prevent them from being satisfactory. An upper limit to ΔV would appear to be set by assuming that ΔS is constant.

If c can be estimated from theory, A can be obtained from the relation $T_0/Ac = (dT/dP)_{P=0}$. Now c can be obtained from Grüneisen's ratio, γ, for the solid through the relation $c = (6\gamma + 1)/(6\gamma - 2)$. Gilvarry (1957) has estimated c to be 1.9 for iron. Al'tshuler *et al.* (1958) have measured Grüneisen's ratio of iron at high pressures using shock-wave techniques; their data lead to $c = 1.4$ (see also Knopoff and MacDonald, 1960).

Taking $\Delta V = 0.31$ cm^3/mole at low pressure, and adjusting ΔS and A to fit Strong's experimental slope of the fusion curve, we obtain at the inner core boundary ($P = 3.2$ megabars) $\Delta V = 0.18$ cm^3/mole if $c = 1.9$, and $\Delta V = 0.11$ cm^3/mole if $c = 1.4$. Since the density of the core is on the order of 10 gm/cm^3 and hence $V \approx 5$ cm^3/mole, these results imply volume changes of 2–4 per cent in the core. Most of the approximations that have been made tend to make this estimate too large, but, nevertheless, a figure amounting to a few per cent is indicated. A further increase in the mean density of the inner core results from compression. Between its boundary and the center of the earth, the pressure rises by about 0.4 megabar. The bulk modulus of the inner core is unlikely to be very different from that at the base of the outer core, about 10 megabars for all modern density distributions, so the increase in density resulting from compression is about 4 per cent. This would be partially offset by a rise in temperature, but both the thermal expansion and thermal gradient are probably small at these great depths. The mean density of the inner core is about 1 per cent greater than the density at its boundary because of compression.

The foregoing estimate of the jump in density at the boundary of the inner core as 5 per cent or less depends on the assumption that the composition does not change appreciably. In view of the known tendency of solid solutions to be more restricted than liquid solutions, this assumption is perhaps more likely to be wrong than right. The segregation coefficient of nickel is probably near unity, and the effect of nickel on the density jump is probably not great. Carbon tends to be concentrated in the liquid as long as its content is not too high. At ordinary pressure, sulfur is almost insoluble in solid iron or nickel, and silicon has restricted solubility in the γ phase. Unless present as carbides, sulfides, or silicides, these elements are probably much less abundant in the inner core than in the outer core. That is, the inner core would have properties closer to pure iron than would the outer core. On the basis of shock compressions of metals, Knopoff and MacDonald (1960) and Birch (1961*a*) have concluded that the outer core is roughly 15 per cent less dense than iron at the same pressure and temperature. If the inner core is essentially iron-nickel, a jump in density of about 15 per cent would thus be anticipated. Adding this to the volume change on solidification, we find roughly 20 per cent as an upper limit to the density jump.

There is a further shred of seismic evidence bearing on this point; it follows from a calculation by Elsasser (1950), corrected by Birch (1952), and depends on the convergence of the properties of solids and liquids at high pressures. One would expect that the value of ϕ at the base of the liquid outer core is a good approximation to K_S/ρ in the solid inner core if the compositions are the same. With this assumption, Poisson's ratio, σ, can be obtained from the velocities. Taking 10.11 km/sec for the velocity at the base of the outer core and 11.1 km/sec for the velocity in the inner core, we find $\sigma = 0.43$. This value is possible for a solid, but it is unusually high, implying comparatively low rigidity. For most solids, σ is between 0.25 and 0.33. But now let us assume that the inner core contains a smaller percentage of light elements than the outer core and that the effect of this is to increase the density by 20 per cent and leave the bulk modulus unaffected. This is simply Bullen's compressibility-pressure hypothesis (Bullen, 1949). With this change, we find $\sigma = 0.35$ in the inner core, a reasonable value. Birch (1952) used a theoretical equation of state for iron to obtain K_S/ρ, and found $\sigma = 0.37$–0.38. The good agreement between these values is doubtless fortuitous, and the result, although suggestive, should not be weighted heavily because of the uncertainties in the seismic data. The negative result of Elsasser's calculation, however, suggests that chemical identity of the outer and inner cores is not compatible with present seismic data.

During the examination of a number of models of the earth's density variation, calculations of densities in the outer core from the Williamson-Adams relation were made in some fifteen cases. Densities at the outer margin of the core ranged from 7.77 to 10.17 gm/cm^3, a far wider range than is permitted in the earth by other considerations. In all cases the rise in density in the outer core was between 2.0 and 2.4 gm/cm^3. Of more significance is the range in z ($=I/MR^2$) encountered in the cases of the two extreme values of the density jump at the boundary of the inner core. In all these calculations the inner core is assumed to be a uniform sphere. Its moment of inertia is so small that this makes little difference. If the density jump is zero, z is found to lie between 0.387 and 0.390; for a jump of 20 per cent, z lies between 0.383 and 0.387. A mean density of around 20 gm/cm^3 in the inner core is required to bring z down to 0.37. It seems clear that z for the core must lie between 0.38 and 0.39. An increase in density throughout the mantle of 0.02 gm/cm^3 is required to reduce z in the core by 0.01.

The ability to limit the value of z for the core places some rather severe limits on the density variation in the mantle. As we have already seen, the upper mantle is inhomogeneous and the Williamson-Adams equation cannot be used. The principal question in this region is how much increase in density between the top of the mantle and a depth of 1,000 km. is required in order to satisfy the requirements set on the value of z for the core. This in turn depends on the density assumed at the top of the mantle, and an important task is to attempt to limit this quantity.

It is assumed that equation (4) holds in the lower mantle. The volumetric thermal expansion is assumed to decrease from 18×10^{-6}/deg to 8×10^{-6}/deg from the top to the bottom of this layer, following Birch's (1952) discussion. The decrease is assumed linear in the calculations; this is not strictly correct, but the effect of a moderate superadiabatic gradient is fairly small and the approximation adequate. The region of vanishing velocity gradient near the core is treated in the same way as the remainder of the lower mantle. Alternative treatments would make little difference because of the restricted extent of this layer.

In the model of the density variation discussed by Birch (1961*a*, *b*), it is assumed that the density depends on the *P* velocity through the empirical relation

$$\rho = A(m) + BV_P \,. \tag{7}$$

Here m denotes the mean atomic weight. In ordinary rocks m is most influenced by the content of iron and, to a lesser extent, by titanium and calcium. Potassium has little influence in the basic rocks because of its relatively low concentration. The constant B is well determined by a large number of measurements on samples with m close to 21. For other values of m the sampling is less complete, and the measurements do not rule out the possibility that B depends on m. A very strong correlation between V_P and ρ is established at $m \approx 21$, however.

A second limitation of the model is its inability to deal with the temperature dependence of velocity and density. This is inevitable because it is based almost entirely on room-temperature data. It is unlikely that $(\partial V_P/\partial T)_\rho$ is zero, although it may be small in most cases. But this implies that both $A(m)$ and B may be functions of temperature.

As is common with empirical models, this one does not predict the densities of certain compounds accurately. The most conspicuous failures are for those high in calcium.

Birch (1961*a*) has discussed all these weaknesses of his model. From many points of view, they are more than compensated by its great strength, which is an explicit recognition of the fact that density and *P* velocity are closely related. It is very unlikely that large changes in density are unaccompanied by large changes in velocity. Although the correlation between velocity and density is not expected to be perfect, it should be close.

Birch (1961*a*) calculated a density curve for the mantle using equation (7). The curve is very close to some of those discussed below. No such definite model is used here because the main aim of this investigation is to examine the differences in density arising from different models. Another recent study of the earth's density variation, also using a high-speed computer, is that of Bullard (1957).

In the following calculations, allowance for the mass and moment of inertia of the crust is the same as that used by Bullen (1940). The mass of the crust is taken to be 4.7×10^{25} gm., and its moment of inertia 1.24×10^{43} gm. cm.^{2}. Allowable changes in these quantities are unlikely to make an important difference in the ensuing calculations. The density is assumed to be constant between the top of the mantle and 150 km. depth. The decrease in velocity in this range of depths implies a decrease in density of about 0.1 gm/cm^3 on Birch's model with constant m. A change in m of about 0.5 would account for the observed velocities at constant density. Such a change cannot be ruled out, and in fact an uncertainty of about this is inherent in Birch's approach. Besides, the details of the distribution of density in a thin shell have little effect on the mass and moment of inertia of that shell. The assumption of a constant density in this region is convenient, since it is immediately comparable with the estimates of densities of subcrustal material afforded by certain petrological observations.

In the transition zone in the mantle, which is currently placed at depths between about 200 and 1,000 km. by seismologists, some simple, geometrical models of the density variations have been examined. These are two broken-line models and one smooth sigmoid variation. Breaks in slope in two of the models are not intended to have any real significance or to bear any close resemblance to the seismic data, which themselves are subject to modification in this region. The models are designed to illustrate in a very simple way the differences in assumptions required in order to achieve a given increase in density between depths of 200 and 1,000 km. in the cases in which the density rises rapidly at shallow depths and then more slowly, and the converse. These models are compatible with the restrictions on the value of z in the core discussed above.

For a density curve of a given general shape the two remaining adjustable parameters are the densities at the top and at the base of the upper mantle. These two parameters interact with each other, and the situation can be simplified if the permissible range of density immediately beneath the crust can be set. There are two approaches to this problem, one through meteorites and one through measurements on rocks supposed to be of profound origin.

In the meteorite model of the upper mantle, an "eclogite norm" can be calculated from the average silicate phase of chondrites of Urey and Craig (1953). In this procedure, determinacy is secured by assuming that all the alkalis are in jadeite, the remaining alumina is in a lime-free garnet, and the remaining major oxides are in olivine and pyroxenes. The ratio of FeO to MgO is assumed to be the same in all the mafic minerals; this is not actually true for coexisting pyroxenes and garnets, but no other course is available. The resulting norm is, in weight per cent and neglecting accessories, 17.3 omphacite (jadeite plus diopside-hedenbergite), 40.2 hypersthene, 38.9 olivine, and 3.5 garnet. The density of such an aggregate, with appropriate iron: magnesium ratio, is about 3.45 gm/cm^3. The partition of iron to the garnet can have little influence because of the small amount of garnet present; an equally plausible norm could be calculated assigning all the alumina to the pyroxene (Boyd and England, 1960; Clark, Schairer, and De Neufville, 1962). This alternative would slightly reduce the average density. Allowing for 1 per cent compression between the surface and the top of the mantle, we find a density close to 3.5 gm/cm^3 for the outermost part of the mantle on this model.

The other source of information about the outermost part of the mantle is through the study of materials that are presumed to have been brought up from great depth. These are principally ultra-basic inclusions in basaltic rocks, and kimberlites and their cognate xenoliths.

A careful study of basic inclusions in basalts has been made by Ross *et al.* (1954). The main minerals are olivine, enstatite, and a chromian augite, with accessory chromian spinel. The mean density of the analyzed olivines is 3.33 gm/cm^3, of the enstatites 3.29 gm/cm^3, and of the augite 3.31 gm/cm^3. The alumina content of the pyroxenes in the inclusions ranges up to more than 5 per cent in the enstatites and more than 6 per cent in the augites. This suggests that these phases truly were formed at high pressures, and it is compatible with origin in the mantle (Boyd and England, 1960; Clark, Schairer, and De Neufville, 1962).

The work of Ross *et al.* (1954) has been cited as evidence that the mantle is a uniform dunite or peridotite. This is not the case. While it is true that the individual minerals (olivine, enstatite, etc.) are rather uniform in composition, the proportions of these phases in the inclusions vary widely. The content of olivine, for example, ranges from virtually 100 per cent to nearly zero. The evidence here is for heterogeneity of source material, not homogeneity.

Kimberlites and, to a lesser extent, their inclusions, of which the most interesting are eclogites, are generally altered and hydrated, so little can be learned about the mantle by direct examination of these rocks. The kimberlites are also extremely basic rocks; even with most of the iron tied up in magnetite, they contain normative Ca_2SiO_4 (Nockolds, 1954). After converting the Fe_2O_3 to FeO, the FeO/(FeO + MgO) ratio is found to be 0.14 in the average "basaltic" kimberlite of Nockolds and 0.18 in his average "micaceous" kimberlite. Olivine with these iron: magnesium ratios would have a density of about 3.4 gm/cm^3, and, since this would be the dominant mineral, it may be taken as a fair indication of the density of the rock itself. Garnet would raise this value, but pyroxenes would reduce it.

A number of measurements of the densities of eclogites from kimberlites are given by Williams (1932), and a few other values have been obtained by Birch (1960). Most are less than 3.4 gm/cm^3, although Williams reports one value of 3.59 gm/cm^3. Some, but not all, of the samples are altered, and this may be a partial cause of some of the low densities. The iron: magnesium ratio is also quite variable in the analyses collected by Williams. It appears that eclogites can have densities as high as 3.6 gm/cm^3, but it also appears that such values are unusual.

A difficulty in attempting to infer mantle properties from kimberlites and from the various types of inclusions is that the specimens are at best *from* the mantle, not *in* the mantle. Hence they may not represent the material that remains beneath the crust. Also, they may have been altered on their way to the surface. This must be borne in mind when assessing the reliability of conclusions derived from such data.

This petrological evidence suggests that the density beneath the crust is higher than 3.3 gm/cm^3 and that it is unlikely to be as high as 3.6 gm/cm^3. Correction for compression would increase these values by about 1 per cent. For the sake of a more or less definite model, it will be assumed that the density at the top of the mantle is between 3.35 and 3.5 gm/cm^3.

Models of the density variation in the upper mantle that satisfy all the conditions imposed have a number of features in common. The rise in density in this layer is not more than 1.2 gm/cm^3, nor can it be less than 0.8 gm/cm^3 unless the density rises rapidly immediately beneath the crust. The latter alternative seems improbable in view of the low-velocity layer. All the acceptable models lead to densities between 4.47

gm/cm³ and 4.62 gm/cm³ at a depth of 1,000 km. The foregoing results follow if the satellite-determined value of the earth's moment of inertia is accepted. The moment of inertia adopted by Bullen would require the density throughout the mantle to be higher by 0.07 gm/cm³. The increase in density in the upper mantle could hardly be less than 1.0 gm/cm³ with the higher total inertia.

The permissible range in density in the upper mantle is fairly well indicated by the curves in figure 3. The different models differ among themselves by up to 0.2–0.3 gm/cm³. Judging by these models, the density is least well determined at shallow depths; the range at 1,000

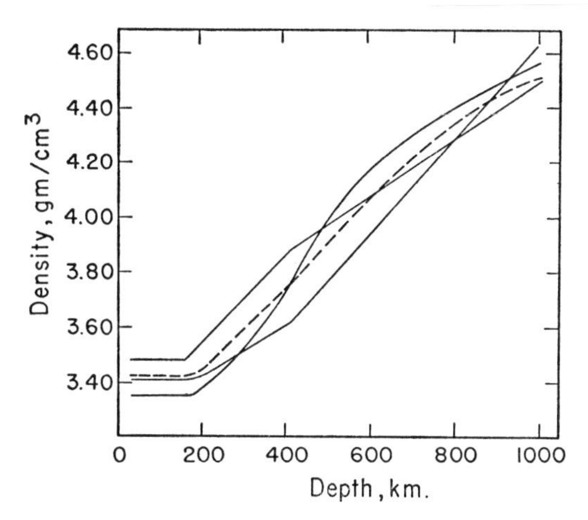

FIG. 3.—Some models of the density variation in the upper mantle. The curve adopted for future discussion is shown as a dashed line.

km. is only 0.15 gm/cm³, as stated above. In ensuing discussions it will be convenient to adopt a definite density curve for the upper mantle, and the dashed line in figure 3 represents the one chosen. It is an intermediate case but is otherwise of no special significance.

With this intermediate curve adopted in the outer mantle, it is a straightforward matter to investigate the influence of a superadiabatic gradient in the lower mantle. An upper limit to the possible magnitude of τ can be set by consideration of the temperatures implied. Most modern attacks on this problem lead to temperatures considerably less than 10,000° C. at the base of the mantle (see, for example, the summary by Gilvarry, 1957), and Birch (1952) found from his analysis of seismic data that a temperature of 10,000° C. was improbably high. Now if $\tau = 2$ deg/km, the total thermal gradient must be greater by a few tenths of a degree per kilometer, which implies that the temperature rises nearly 5,000° C. in the lower mantle. This, in turn, leads to temperatures of 6,000°–8,000° C. at the base of the mantle. These values are acceptable. A superadiabatic gradient of 3 deg/km leads to temperatures that are not acceptable, and 2 deg/km will be taken as a round upper limit to τ. It is assumed that τ is not negative.

The chief effect of a positive value of τ is to reduce the density gradient in the lower mantle. The adopted densities in the upper mantle require adjustment in order to maintain the value of z for the core within its prescribed limits. The adjustment was made by increasing the density in the upper mantle by a uniform amount at all depths; the increase

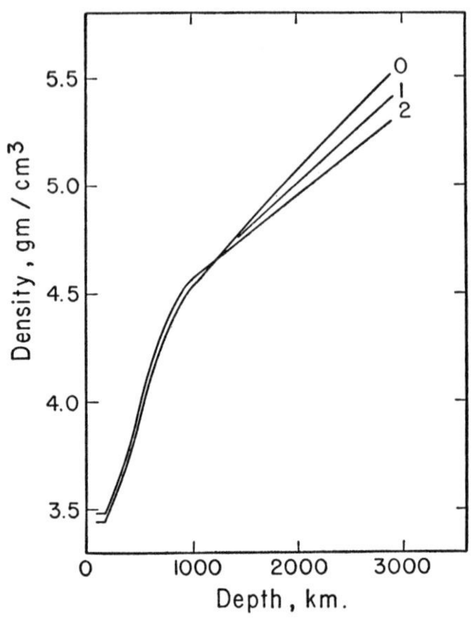

Fig. 4.—The density variation in the mantle. The different curves are for different values of τ, which are indicated by the numbers, representing deg/km, beside them.

required was 0.04 gm/cm³ when $\tau = 2$ deg/km. The density is affected most at the base of the mantle, where the value for $\tau = 0$ is greater than that for $\tau = 2$ deg/km by 0.22 gm/cm³. The effect of a superadiabatic gradient in the lower mantle on the densities is illustrated in figure 4.

Density curves for the entire earth are shown in figure 5. The adopted curve in the outer mantle is adjusted to be compatible with a superadiabatic gradient of 1 deg/km in the lower mantle. Hence this too is an intermediate case. The value of z for the core is 0.387. Two curves are shown in figure 5; the heavy solid curve was calculated using the moment of inertia determined from satellite orbits, and the light dashed curve was obtained from the higher inertia used by Bullen. The density in the mantle is about 1–2 per cent higher, and in the core about 3 per cent lower, when the new inertia is adopted.

The densities given by Bullen's model A (Bullen, 1947) are also shown in figure 5 for comparison. The most important ways in which model A differs from the present models are in the lower density assumed at the top of the mantle (3.32 gm/cm³) and the assumption of adiabatic conditions in the lower mantle. The latter difference between the models is primarily responsible for the higher density in the outer core in

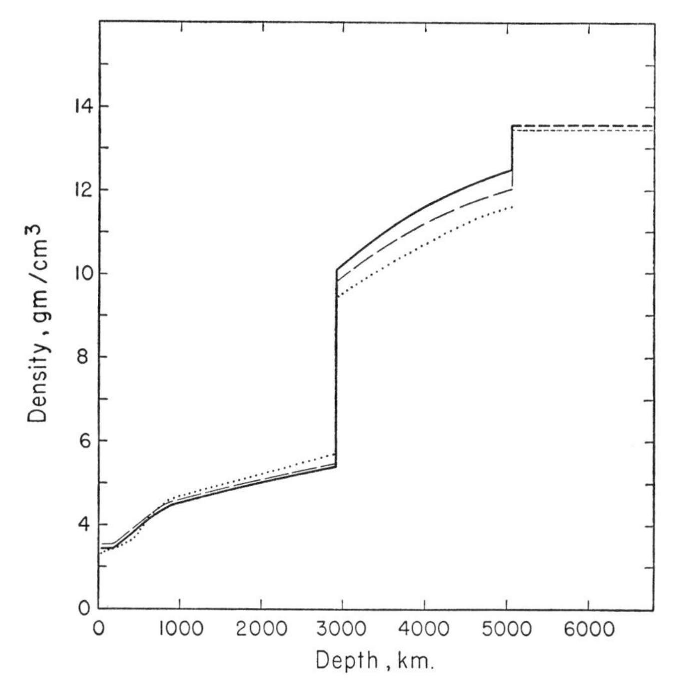

FIG. 5.—An "intermediate" model of the density in the earth. The heavy curve is calculated with the moment of inertia from satellite data. The light dashed curve is calculated using the same moment of inertia as Bullen. In both cases τ = 1 deg/km in the lower mantle. The dots represent Bullen's model A.

the present case. The densities in the inner core in model A are arbitrary and devoid of chemical or physical significance; hence they are not shown in figure 5.

The uncertainty in the density distributions can be estimated in a limited sense from the differences between the models examined. Additional uncertainty in the densities arises from uncertainties in the mass and moment of inertia of the earth, the seismic velocities, the various assumptions about adiabaticity and allowance for departures therefrom, and the densities or changes in density at certain levels in the earth. These sources of error are ignored in what follows, and, in addition, it must be recognized that occasional excursions of the density beyond the stated limits cannot be excluded, particularly in the upper mantle.

At the top of the mantle the density is restricted by assumption to within about 2 per cent. The uncertainty increases to about 4 per cent in the upper mantle and declines to 2 per cent or so at a depth of 1,000 km. It gradually rises in the lower mantle, chiefly because no close limits can be set on the thermal gradient, and it is close to 3 per cent at the base of the mantle. The uncertainty in the densities in the core also results in large part from the wide latitude in the value of τ permitted by present data. This alone implies that the densities in the outer core cannot be known to better than 2 per cent, and a possible error of 1–2 per cent is present, in addition, because of differences between the various models of the outer mantle. In the inner core, of course, the density is uncertain by more than 10 per cent because the density jump at the margin cannot be specified more closely than that.

There are several places where tighter controls on assumptions are both reasonable to expect in the foreseeable future and particularly important in order to narrow the uncertainties in the density distribution. The most obvious of these is the top of the mantle. More sophisticated analyses of petrogenetic relationships may permit closer limits on density in this region to be set, and there is the further possibility that direct sampling may become feasible through deep drilling. In the upper mantle more definite mineralogical models with their associated implications about phase changes will evidently help in limiting the range of density. It is perhaps less obvious that a close upper bound on the central density would also be very valuable. This would limit the allowable mass of the core and hence the mass of the mantle, and thus an upper limit to the thermal gradient in the lower mantle could possibly be established.

The density at a depth of 1,000 km. has been found to be close to 4.55 gm/cm^3, and ϕ at the same level is about 77 (km/sec)^2 according to the seismic data. We must inquire as to what materials could have these elastic properties at the relevant pressure (about 400 kb.). Actually, a somewhat higher elasticity and density are required at room temperature in order to make allowance for the high, although unknown, temperature in the mantle.

The compression curve of stishovite can be estimated from the shock compression of quartz (Wackerle, 1962). The data are shown in figure 6, and it is noteworthy that there is a sharp break in the curve near 400 kb. and that at somewhat higher pressures the points lie on a nearly straight line pointing toward the density of stishovite at atmospheric pressure (Chao *et al.*, 1962). The interpretation of these results in terms of a transition to stishovite under shock conditions is certainly tempting;

they imply a density of around 4.5 gm/cm^3 at 400 kb. and a very high mean value of K/ρ, about 125 (km/sec)^2.

The elastic parameters of other oxides at this pressure can be estimated with the aid of Birch's (1952) extension of Murnaghan's theory of finite strain. This yields the relations

$$P = 3K_0 f(1+2f)^{5/2}(1-2\xi f), \tag{8}$$

and

$$K/\rho = (K_0/\rho_0)(1+2f)[1+7f-2\xi f(2+9f)], \tag{9}$$

where the subscript zeros refer to zero pressure, and

$$f = \frac{[(V_0/V)^{2/3}-1]}{2} = \frac{[(\rho/\rho_0)^{2/3}-1]}{2}. \tag{10}$$

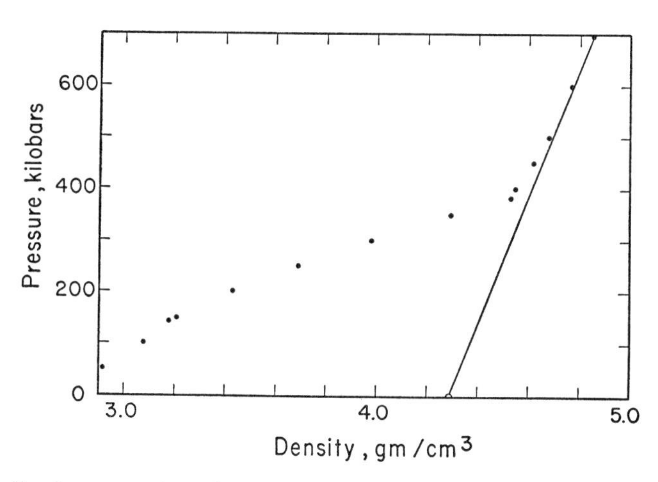

FIG. 6.—Shock compression of quartz. The solid points represent smoothed data from Wackerle (1962), and the open circle is the zero-pressure density of stishovite according to Chao *et al.* (1962).

Birch examined the magnitude of the dimensionless parameter ξ for a number of substances, some of which had been measured to very large compressions. He concluded that ξ was small, and it is taken to be zero below.

Values of density and K/ρ for several oxides at 400 kb. are collected in table 1, together with these parameters for stishovite and the mantle. In the calculations involving the Birch-Murnaghan theory, isothermal bulk moduli were used. The value of K/ρ for the mantle is ϕ, the adiabatic quantity, and for stishovite it is along the Hugoniot curve, which is neither isothermal nor adiabatic; Wackerle (1962) calculated that the temperatures were in the range 2,000°–3,000° C. at the pressures at which stishovite appeared. Hence it is not always the same quantity that is entered in the last column in table 1, but the differences, which may amount to several per cent, need not concern us here. The low-pressure compressibilities of Bridgman (1949) for Al_2O_3, MgO, and Fe_3O_4, and of Weir (1956) for CaO, and the X-ray diffraction-determined densities of Swanson *et al.* (1953) for Al_2O_3, CaO, and MgO, and of Tombs and Rooksby (1951) for Fe_3O_4 were used in calculating the quantities in table 1.[3]

From inspection of table 1 it seems clear that a mixture of these oxides, which are considerably more abundant than any of the others in the chondrites (Urey and Craig, 1953; Edwards and Urey, 1955), can have the properties required in the deep mantle. An appreciable content of stishovite will raise the average value of K/ρ to amply high values, and the iron oxides can evidently provide the required high density.

TABLE 1

PARAMETERS OF SOME OXIDES AT 400 KB.

	ρ gm/cm^3	K/ρ (km/sec)^2
SiO_2 (stishovite)	4.6	125
Al_2O_3	4.5	95
Fe_3O_4	6.1	50
MgO	4.2_5	75
CaO	4.2	60
Mantle (unknown temperature)	4.5_5	77

Present data cannot yield more than this qualitative generalization because there are no measurements of the compressibility of wüstite, which is a far more likely component than magnetite in the lower mantle. Wüstite (strictly $Fe_{0.953}O$) has a density of 5.745 at zero pressure (Willis and Rooksby, 1953) compared with 5.200 for magnetite; the values of K/ρ are probably also roughly comparable.

It is worth noting that the three oxides in table 1 that have roughly the same mean atomic weight—stishovite, corundum, and periclase—exhibit a relation between K/ρ and valence of the cation that is remarkably regular, in view of the ways in which the data were obtained. That is, the increase in K/ρ on passing from periclase to corundum is nearly the same as on passing from corundum to stishovite. It is of course not clear just how much "regularity" is to be expected, but in any event there is no indication that there is anything seriously wrong with the procedure used to obtain the value of K/ρ for stishovite.

[3] The density of CaO may be underestimated by 10 per cent or so because of neglect of a possible phase transition from the NaCl to the CsCl structure. This was pointed out privately by W. S. Fyfe.

The demonstration that a mixture of oxides such as is discussed here could have the seismic properties and density of the lower mantle does not by itself make such an assemblage an attractive model for the mantle. It must also be shown that such a mixture is thermodynamically stable. A mixture of oxides, including stishovite, is considerably denser than are ordinary silicates. This is largely the result of the closer packing achieved when the coordination number of silicon increases from 4 to 6. Periclase plus stishovite is 21 per cent denser than forsterite, and Wentorf (1962) estimates that the $P\Delta V$ energy of the reaction $Mg_2SiO_4 = 2MgO + SiO_2$ (stishovite) is 21 kcal/mole at 130 kb. MacDonald (1962) and Ringwood (1962) have tried to estimate the pressures at which decomposition to the oxides would take place; their results indicate pressure of a few hundred kilobars at 1,000°–2,000° C., but this is highly uncertain, since the field of stability of stishovite has not yet been delineated. Recent syntheses (Sclar *et al.*, 1962) suggest that stishovite is the stable polymorph of silica at lower pressures than had previously been supposed, and this may indicate that the pressures estimated by MacDonald and Ringwood will require downward revision. It seems highly plausible to suppose that a mixture of oxides is stable in the lower mantle relative to ordinary silicates.

The inversion of olivine to a spinel structure does not affect this conclusion. In forsterite, the volume change accompanying this inversion is variously estimated from 4.5 to 9 per cent (Dachille and Roy, 1960; Ringwood, 1962). The effect of such an inversion, in which silicon remains in fourfold coordination, would be to postpone the decomposition of the oxides to higher pressures. Both Ringwood and MacDonald have taken account of this transition in their estimates of the stability of the assemblage of oxides.

Ringwood (1962) has made an alternate proposal. He considers that magnesiowüstite would not be stable with stishovite but that a compound with metasilicate composition and the corundum structure would be formed. This would also have silicon in sixfold coordination, and, according to Ringwood's estimate, its density is nearly the same as that of a mixture of the constituent oxides. It seems likely that the value of K/ρ would also be very high for this compound (cf. the value for corundum in table 1) and that it, too, would have seismic properties appropriate to the lower mantle. The question of whether this phase or the chemically equivalent mixture of oxides is present in the lower mantle will have to be settled by chemical arguments, since the geophysical effects apparently would be very small. Other dense silicate structures may also be stable, but the arguments of Ringwood and MacDonald show that ordinary silicates are unstable in the lower mantle.

Because of the lack of data on the compressibility of wüstite, no direct estimate of the iron content of the lower mantle can be obtained from the seismic results. Birch's model, equation (7), can give a rough estimate of this through a plot of the adopted mean density as a function of *P* velocity. This is shown in figure 7; the adopted solution is appropriate to a nearly constant mean atomic weight of about 22.5 (i.e., a rather iron-poor composition). This is a very rough estimate, since the effect of temperature has been neglected. Further work is clearly required in order to limit the iron content of the mantle more closely.

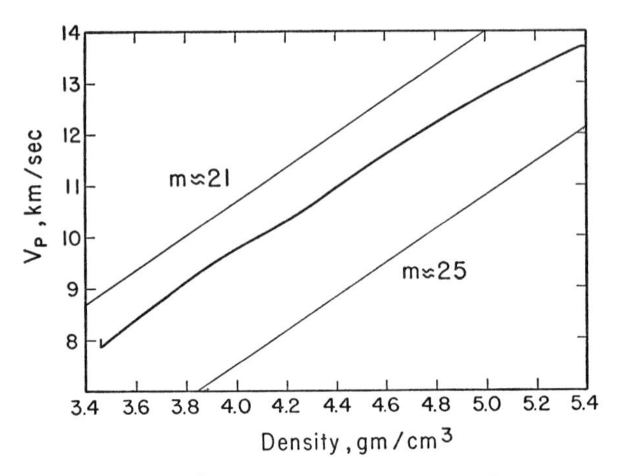

FIG. 7.—Velocity-density plot for the mantle. Lines corresponding to mean atomic weights of 21 and 25, according to Birch's model, are also shown.

THE MELTING CURVE IN THE MANTLE

Recognition of the fact that phase changes occur in the outer mantle means that the question of the melting curve in the mantle must be re-examined. The usual procedure in the past has been either to extrapolate Simon's equation to the core or to attempt to derive the melting curve from the seismic velocities. Neither of these procedures proves to be valid when inhomogeneity is present.

The term "melting point" is not always carefully defined in discussions of a complex multicomponent system such as the mantle. It will be used here to denote the liquidus, or highest temperature at which crystals are stable at a given pressure. Most of the following discussion is concerned with the behavior of an absolute minimum point on the liquidus. Here the solidus and liquidus coincide, and in a given system there can be no liquid at a lower temperature, at the pressure under consideration.

There are a number of reasons why the melting curve in the mantle is of importance. The two most obvious are that it sets an upper limit to present temperatures in the mantle and that it provides a fairly definite and, on some models, appealing estimate of the "initial" temperatures in the earth. But there are also many properties of solids, such as strength, ductility, diffusion coefficients, and thermal diffusivity, which are very sensitive to the proximity of the actual temperatures to the melting points of the individual phases. Laws of corresponding states correlating these properties to a variety of materials can be found by dividing the temperatures by the melting points.

The simplest example of the effect of a transition on the melting curve is in the case of a one-component system. The slope of the melting curve, dT/dP, increases discontinuously at the triple point between two solid phases and liquid. In multicomponent systems the transition line between the solid phases becomes a zone, and the slope of the liquidus increases in small discontinuous steps that approximate a gradual change in passing through this zone. This is opposite to the tendency of melting curves to become less steep at high pressures in the absence of a transition in the solid. Such a tendency is expressed in, and in fact is the basis for, the Simon equation.

Uffen (1952) and Gilvarry (1957) have presented melting curves for the mantle derived from seismic data. Connection between the velocities and the melting temperature was made through extension of Lindemann's melting criterion (see, for example, Richtmeyer and Kennard, 1947, p. 430). This approach is of course open to question because the mantle certainly does not behave as a one-component system. There is also doubt as to whether Lindemann's criterion is applicable to ionic crystals (Clark, 1958), perhaps for the following reasons. Lennard-Jones and Devonshire (1939) showed that it was possible to derive the Lindemann melting criterion from their order-disorder theory of melting. But their theory does not take account of "holes," or vacant sites in the liquid, and yet it is known that these are an important part of the constitution of molten salts (see, for example, Bockris and Richards, 1957). The most direct bit of evidence for this is the volume change on fusion, which amounts to only a few per cent for most metals but which exceeds 20 per cent for the alkali halides. Now it cannot be argued that a theory of liquids that does take account of holes, such as those reviewed by Rowlinson and Curtiss (1951), could not also lead to the Lindemann melting criterion. But the neglect of holes is at least a tempting explanation of the fact that Gilvarry's theory, which predicted the Simon coefficients of the alkali metals with good precision, should fail completely for sodium chloride (Gilvarry, 1956; Clark, 1958). The volume change on fusion is between 2 and 3 per cent for the alkali metals other than lithium (Bridgman, 1952, p. 212), indicating a very small number of holes in the liquid.

Silicates are at least partly ionic compounds, and any theory of melting based on the Lindemann law should be applied to them with reservations. Besides, the relations used by Uffen and Gilvarry to calculate the melting curve are inapplicable in inhomogeneous media. As both these authors recognized, their equations are inapplicable in cases of chemical inhomogeneity; this follows from the fact that they cast their expressions in terms of ratios in order to cancel out the molecular weight of the material. This procedure evidently assumes chemical homogeneity.

Physical homogeneity is also necessary if these methods are to apply. Uffen (1952) gives two expressions for the melting curve. One involves ϕ and the other a combination of the individual velocities. Gilvarry's expression involves both ϕ and a function of Poisson's ratio, which is expected to vary only slightly. At a sharp transition in a one-component system the density changes discontinuously, and by Birch's empirical relation, equation (7), at least one velocity should do the same. Any other behavior would be purely coincidental in view of Bridgman's (1952, p. 247) observation that at a transition the change in compressibility could be positive, negative, or zero. From this it follows that, in general, Uffen's or Gilvarry's expressions would predict a discontinuity in melting point at a sharp transition in a one-component system rather than a change in slope. This is thermodynamically impossible. For a theory to succeed in a multicomponent system when it fails in a system of one component requires another improbable coincidence, and these methods seem to be a most unreliable way to approach the problem of the melting curve in the mantle.

It is possible to arrive at an estimate of the melting curve in the mantle purely from empirical and thermodynamic grounds. Such an approach necessarily involves extrapolations to high pressures and temperatures. This inevitably introduces uncertainty, and it is difficult to determine a priori just how great this is. Following a description of the procedure and some of its consequences, this question will be reopened and the problem of how seriously the results should be taken will be re-examined.

The method followed here takes only approximate account of the fact that the mantle is a multicomponent system, in that it uses the ordinary, simple form of Clapeyron's equation and also Simon's equation. The validity of the latter has never been demonstrated except in the case of a one-component system, although it should give a good approximation to the behavior of a minimum melting point in a multicomponent system. The composition of this minimum is, of course, not constant in general along the fusion curve. Clapeyron's equation is readily generalized to systems of more than one component (see, for example, Morey, 1936), but at present the data required by the generalization simply are not available. The following treatment should be regarded as leading to a zeroth-order, rather than first-order, approximation to the melting curve in the mantle. Some idea of the possible importance of the effect of the transition zone can be gained from it, however.

It is assumed that in the absence of phase changes and chemical heterogeneity, Simon's equation describes the melting curve, and the Birch-Murnaghan theory, equations (8) and (10), with $\xi = 0$ provides pressure-volume relations for the solids. Birch's original derivation of equation (8) assumed isothermal compression. It is easy to derive formally identical relations that apply to adiabatic conditions (Clark, 1959), but the constants in equation (8) are not so well known in this case. In the following argument, equation (8) will be used along the melting curve, which is neither isothermal nor adiabatic. These distinctions are unimportant at the present level of approximation, and isothermal constants will be used.

The following procedure amounts essentially to setting up a hypothetical homogeneous mantle and then correcting for the effects of the transition zone. The Simon equation gives $\Delta V/\Delta S$ along the fusion curve according to equation (6). Since ΔS can be limited by arguments like those used in the case of the melting curve of iron, ΔV can be calculated on various models. The volume of the solid is obtained from equations (8) and (10), and when combined with ΔV gives the volume of the liquid, V_L. This roundabout way of calculating V_L seems unnecessary, since in principle the equations of state (8) and (10) are applicable to liquids as well as solids. We know much less about the parameters K_0 and ξ for liquids, however, and it is not clear that equation (8) includes terms of order high enough to express the compression of liquids adequately. Presumably because of the elimination of "holes," many liquids are initially highly compressible, and equation (8), which, with retention of the term in ξ, is still only a third-order expression, may not be capable of representing the large change in compressibility at low pressures. This difficulty is avoided by the present approach to the equation of state of the liquid along the fusion curve, but at the expense of introducing ΔS as an additional parameter to be evaluated.

The transition zone in the mantle implies an anomalous change in entropy as well as in volume with depth. It is assumed that there is a constant slope to the transition region in the pressure-temperature projection and that the anomalous decrease in entropy, ΔS_{tr}, is related to the anomalous decrease in volume, ΔV_{tr}, by Clapeyron's equation at all points within the transition zone and at greater depths. ΔV_{tr} is then equated to the difference between the volume calculated from equations (8) and (10) and the volume deduced from the density curve in the mantle. The volume change on melting is the difference between the volume of the liquid and the volume obtained from the density curve, and the entropy change is the sum of the change assumed on the basis of a model that neglects phase transitions and ΔS_{tr}. The melting curve is then found by numerically integrating dT/dP as obtained from Clapeyron's equation.

This method of calculating the melting curve is strictly applicable only to one-component systems as outlined. A very approximate extension to more complicated systems can be made by assuming that the main effect of additional components is to reduce the melting point (in the sense defined above) and to reduce the initial slope of the melting curve. This reduction in slope at low pressures is to be expected on thermodynamic grounds. The volumes of most solutions are fairly closely approximated by linear functions of mole fractions. The entropies of these solutions are rarely, if ever, described so simply. A purely configurational term, the ideal entropy of mixing, increases the partial molar entropies of the components, but no such term affects the partial molar volumes. In most chemical systems the extent of solution in the liquid state exceeds that in the solid state, and the result is that there is a minimum on the liquidus. If the solid solubility exceeded the liquid solubility, a maximum point on the liquidus would result, and, if the two solubilities are nearly equal, the phase diagram is analogous to that showing the melting relations of the plagioclase feldspars, with neither a maximum nor a minimum point on the liquidus in the interior of the system.

The existence of a minimum on the liquidus implies a greater entropy of mixing in the liquid than in the solid, and a reduction in the slope of the melting curve of the thermal minimum in the mixture compared with that of the relevant end members, because, roughly, $dT/dP = \Delta V/\Delta S$. Since all known liquidus diagrams of silicate systems show minima rather than maxima, it is inferred that, in general, the slopes of melting curves in multicomponent systems will be less than those in systems that behave as if there were only one component.

Data on the effect of pressure on the liquidus are rare in multicomponent silicate systems. Segnit and Kennedy (1961) have investigated a composition in the system muscovite-silica-water and obtained an initial slope of 6.5 deg/kb. In their studies of the stability of pyrope, Boyd and England (1962) have fixed the temperature and pressure at which the garnet first appears on the liquidus. This determines the meant slope of the liquidus of the low-pressure assemblage that is stable on the pyrope composition. The low-pressure behavior is ternary, and the mean slope between 0 and 36 kb. is 6.4 deg/kb. Clark, Schairer, and De Neufville (1962) have studied the minimum in the system diopside-anorthite-silica at 0 and 20 kb. The mean slope in this range of pressure is 6.3 deg/kb. In the last case, the extent of solid solution is much greater at high pressure than at low, which tends to increase the slope of the melting curve. Yoder and Tilley (1962) have determined the slope of the liquidus of natural basalts and eclogites. In the field of stability of basalt the slope is about 6.5 deg/kb, and in the eclogite field about 11 deg/kb. For comparison, the initial slopes of the melting curves of diopside and albite are about 15 deg/kb, and the average slopes up to 20 kb. are 10 to 12 deg/kb (Boyd and England, 1963). In binary systems not involving silicates, Newton, Jayaraman, and Kennedy (1962) have found that the effect of pressure on the eutectic in the system Na-K is less than the effect on the pure metals. Bovenkerk *et al.* (1961) indicate that the melting curve of nickel is steeper than the curve of its eutectic with carbon.

The Simon exponent, c, is somewhat greater than 4 in the cases of diopside and albite. This will be taken as a representative value for pure silicates that melt congruently. The value of c appropriate to the minimum must be at least as large as its value at the end members, since otherwise equation (5) predicts that the melting curves of the minimum and the end members would cross. This is inconsistent with the requirement that the minimum temperature should rise less rapidly with pressure than the melting points of the end members. It also implies, of course, that the minimum would eventually become a maximum, which is most improbable.

For the sake of a definite model, the initial slope of the melting curve will be taken to be 6.5 deg/kb. We assume a melting point of 1,300° K. at zero pressure, and a round value of 4 for c, the Simon exponent. Values of 15 and 25 deg/kb were taken for the slope of the transition zone; this is approximately the range of slopes estimated by Dachille and Roy (1960) for the olivine-spinel transition in forsterite. The slope of the transition involving decomposition to the oxides may be greater than this, but it is unknown and these values must suffice for the present.

The initial bulk modulus of the solid material was taken to be 1.25 megabars, a value appropriate to olivine (Bridgman, 1948). The initial volume in equation (10) was adjusted so that the calculated density of the solid agreed with that given by the heavy curve of figure 6 at a depth of 100 km.

The final parameter to be considered is $\Delta S_f{}^0$, the entropy change on fusion in the absence of a transition. For silicates the heat of fusion is commonly in the neighborhood of 100 cal/gm (420 j/gm). If the molecular weight is 150 (again appropriate to a magnesian olivine) and the melting temperature is 1,300° K., $\Delta S_f{}^0$ is between 40 and 50 j/mole deg. (The molecular weight affects both molar volume and molar entropy in the same way and cancels out in Clapeyron's equation; hence the value used has no significance in the calculations of temperature.) A value of 45 j/mole deg was assigned to ΔS_f^0 at zero pressure.

The arguments about the pressure dependence of ΔS_f^0 given above for iron are applicable with additional force to silicates. These partially ionic liquids may be expected to contain appreciable quantities of holes; as the holes are expelled by pressure, the configurational term in the entropy is decreased. This process takes place in addition to the reduction in the volume change on fusion, which, according to Slater's (1939) argument, should also lower ΔS_f^0. It is, however, very difficult to estimate the magnitude of these effects in the earth. It was rather arbitrarily assumed that ΔS_f^0 decreases linearly with pressure to half its surface value at the base of the mantle.

The entropy change on fusion along the actual fusion curve with the transition zone present, denoted ΔS_f, contains a second term that causes it to decrease with increasing pressure. This is ΔS_{tr}, the entropy change associated with the transition region, which is a linear function of ΔV_{tr}, the corresponding volume change. The latter is the difference between the volumes of the denser, relatively incompressible, material of the lower mantle and the less dense, relatively compressible, material at the top of the mantle. Because of this difference in compressibility, ΔV_{tr} and, hence, ΔS_{tr} decrease with increasing pressure, producing a further drop in ΔS_f, which equals $\Delta S_f^0 + \Delta S_{tr}$.

Melting temperatures calculated by these assumptions are shown in figure 8. The curves have the expected shape, convex toward the depth axis in part of the upper mantle and gently concave toward it in the lower mantle. The melting point at the base of the mantle is 7,350° C. if the slope of the transition region is 15 deg/kb, and 8,100° C. if the slope is 25 deg/kb. Uffen's (1952) melting temperatures are also shown in figure 8. The shape of the curve is similar to the shapes of Jeffreys' velocities, which place the transition zone between 400 and 1,000 km. The abrupt rise at the top of the transition zone and the immediate decrease in melting point gradient on entering the lower mantle are both features that seem implausible on thermodynamic grounds. Furthermore, the method of obtaining these curves insures that they are independent of the slope of the transition zone. The present treatment shows this to be most improbable.

It is convenient to discuss the influence of changing some of the parameters in the expressions in terms of the temperature at the core boundary. Increasing ΔS_f by 50 per cent decreases the calculated temperature at that point by about 1,000° C. If ΔS_f^0 is assumed to be constant and equal to its value at zero pressure, the temperature is found to be about 500° C. less. If ΔS_f^0 is assumed constant and equal to the mean value of the entropy in the case of variable ΔS_f^0 (that is, three-fourths of the zero-pressure value), the temperatures are virtually identical.

Changing the slope of the melting curve at zero pressure proves to make little difference. Increasing this quantity from 6.5 deg/kb to 10 deg/kb affects the temperature at the core boundary by only about 300° C. As shown in figure 8, the two values adopted for the slope of the transition region lead to temperatures that differ by somewhat less than 1,000° C. Either the remaining parameters are fairly well fixed by geophysical

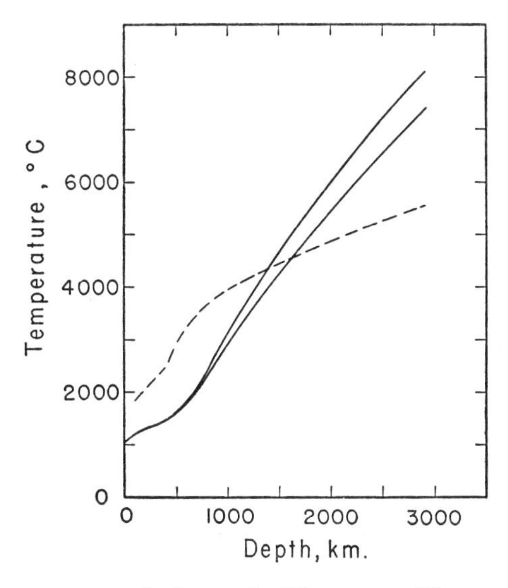

FIG. 8.—Melting temperatures in the mantle. The upper solid curve is for a slope of 25 deg/kb in the transition zone, and the lower is for a slope of 15 deg/kb. The dashed curve shows the melting temperatures obtained by Uffen (1952) from seismic data.

or experimental data or changes in them have little effect on the calculated temperatures. The conclusion is that this model leads to a temperature of about 7,500° C. at the core boundary and that the uncertainty in this figure is less than 2,000° C.

It is also interesting to study the cooling of the mantle on the additional assumption that the initial temperatures equal the melting points as calculated above. Most modern geomagnetic theories require that the thermal gradient in the outer core be adiabatic, and the heat conducted into the mantle down such a gradient proves to be considerable. The simplest and in many ways the most appealing source of this heat is the latent heat released by the crystallization of the inner core. This process will maintain an adiabatic gradient in the outer core only if the temperature at the core-mantle interface is decreasing with time, since the melting-point gradient must exceed the adiabatic on this model. An important result of a study of the cooling of the mantle will be an estimate of the lowest value of thermal conductivity in the lower mantle that permits this condition to be met.

This problem will be approached through the linearized equation of heat conduction with the thermal conductivity assumed to be constant. We must find the temperature in a spherical shell $r_0 \geq r \geq r_i$, with $T = 0$ at $r = r_0$ and constant flux q at the boundary $r = r_i$. The initial temperature is $f(r)$, the temperature given by the melting curve.

If we set

$$u = T + \frac{q r_i^2}{K r_0} - \frac{q r_i^2}{K r}, \tag{11}$$

and $v = ur$, then v must satisfy

$$\frac{\partial^2 v}{\partial r^2} = \frac{1}{\kappa} \frac{\partial v}{\partial t} \tag{12}$$

and

$$v = r f(r) + \frac{q r_i^2 r}{K r_0} - \frac{q r_i^2}{K}, \tag{13}$$

when $t = 0$. Also $v = 0$ when $r = r_0$, and $\partial v / \partial r = v/r$ when $r = r_i$. In these equations K is the thermal conductivity, κ the thermal diffusivity, and t the time. The solution to these equations is

$$v = \Sigma_n A_n (\sin a_n r - \tan a_n r_0 \cos a_n r)\, e^{-a_n^2 \kappa t}, \tag{14}$$

where the a_n are the positive roots of

$$a_n r_i + \tan a_n (r_0 - r_i) = 0, \tag{15}$$

and the A_n are chosen to satisfy equation (13) when $t = 0$.

If the initial temperatures are given as an analytical function of r, standard Fourier methods can be used to evaluate the A_n. But in the present case the initial temperature is defined only in terms of 22 numerical values at 22 distinct radii, as evaluated by the computer. Hence a truncated series of the form given in equation (14) must be fitted by some numerical procedure. The one adopted was to fit a 21-parameter expression to the 22 points by least squares. This was chosen over the nearly equivalent procedure of exactly fitting a 22-parameter expression to the points for the simple reason that it proved to be more convenient with existing programs for the computer. The computed values of the initial temperature were fitted to within 1 degree or so by the least-squares program.

The coefficients A_n cannot be closely determined beyond about A_{11} because the wave lengths of their oscillations approach the distance between the calculated points. For t greater than 10^9 years, these terms are negligible, however, because of the exponential factors. An 11-parameter expression gives virtually identical results at these times, although it does not lead to a good fit to the data at $t = 0$.

The heat flow from the core is equal to its thermal conductivity times the adiabatic gradient. Hide (1956) gives 0.1 cal/cm sec deg for the former quantity but cautions that this figure is uncertain by a factor of 5. Combining it with Birch's (1952) estimate of the adiabatic gradient in the core, we find that the heat flow is $0.15 \times 10^{-9}T$ cal/cm^2 sec, with T in degrees Kelvin.

With Hide's value of the thermal conductivity in the core, the corresponding conductivity in the mantle must be about 0.04 cal/cm sec deg to maintain convection in the core if the initial temperatures are as shown in figure 8. Other models leading to lower temperatures at the core boundary have also been investigated; they require values of thermal conductivity in the mantle that are higher than the figure given above by a few per cent. This is because the thermal gradient in the mantle that is available to carry away the heat from the core suffers a greater proportional reduction than does the adiabatic gradient.

The value of the thermal conductivity deduced from this model is obtained on the assumption that the radioactive generation of heat in the lower mantle is very small. It seems probable on thermal grounds that this is the case, but it cannot be concluded that the radioactive elements are completely absent from the lower mantle. Hence a thermal conductivity between 0.04 cal/cm sec deg and 0.05 cal/cm sec deg may be required. It seems possible to provide this through a combination of lattice conduction and radiative transfer. If, however, future work should require upward revision of the estimated thermal conductivity in the core, the possibility of large-scale movements of material in the mantle would have to be given further consideration.

These models lead to considerable cooling in the lower mantle, the exact amount depending on the thermal conductivity. At a depth of 2,000 km., which is considerably removed from the core-mantle boundary, the cooling in 4.5×10^9 years is about 250° C. if $K = 0.015$ cal/cm sec deg, and about 600° C. if $K = 0.04$ cal/cm sec deg. The presence of deep-seated radioactivity would reduce these figures, and it appears that the resulting thermal contraction could not reduce the circumference of the earth by more than a few tens of kilometers. It also follows that appreciable radioactivity, perhaps up to 10 per cent of that found in the average chondrite, could be accommodated in the lower mantle without causing heating.

We must now return to the question of whether the foregoing model is acceptable. Extrapolation of the fusion curve for iron, assuming Strong's (1959) slope at low pressure and a value of 1.4 for the Simon exponent, leads to a temperature of about 7,500° C. at the boundary of the inner core. This is probably an upper limit to the temperature there, since the effect of alloying elements is to reduce the melting temperature. The increase in temperature through the outer core is somewhat less than 2,000° C. at a mean temperature of 7,000° C. (Birch, 1952), and the melting point of iron at the outer boundary of the core, again from the Simon equation, is about 4,000° C. The melting temperatures given in figure 8 are compatible with the liquidity of the outer core but are incompatible with solidity of the inner core—unless the lower mantle is assumed to be well below its fusion temperature. It seems advisable to reconsider the model before accepting this latter means of reconciling the data.

Of all the assumptions underlying this model, the use of Simon's equation in the way described seems most open to question. The relationship assumed implies that the liquid can never become denser than the extrapolated density of the low-pressure solid phases. We have already seen that a change in the coordination number of silicon from 4 to 6 is probably required to account for the densities of the solid phases in the lower mantle. Such a change in average coordination may also take place in silicate liquids, although spread over a broad, vague range of pressure. This would lead to the volume of the liquid becoming smaller than is allowed by the present model, which would produce a greater downward curvature to the melting curve in the lower mantle. This could reduce the temperature at the core-mantle boundary by a few thousand degrees. Until a satisfactory way of dealing with this possibility is devised, the present approach to the melting curve in the mantle can be considered to be of only limited success. On the other hand, it does point up, in a qualitatively correct way, the important effect that phase changes in the upper mantle have in steepening the melting curve at depths near 500 km.

Conclusions

It is the purpose of the present volume to emphasize unsolved rather than solved problems. Some of the more important questions raised, and not satisfactorily answered in the present chapter, are collected below.

1. What are the magnitudes of the departures from spherical (or more accurately, elliptical) symmetry in the properties of the earth?
2. What is the mean density at the top of the mantle, and what is the range of density?
3. What sequence of phase changes is expected in the upper mantle, and what forms of density variation do they imply?

4. What are the elasticities of magnesiowüstites of various compositions?

5. What is the effect of temperature on various elastic properties as a function of pressure? In particular, what is the temperature coefficient of compressional velocity at constant density?

6. What is the highest allowable density at the center of the earth?

7. What is the equation of state of silicate liquids, especially at very high pressures?

8. What is the thermal conductivity to be expected in the outer core?

9. What is the thermal conductivity to be expected in the lower mantle, taking account of radiative transfer?

ACKNOWLEDGMENTS.—I am indebted to Francis Birch for calling my attention to the shock-wave data for silica and their interpretation and to H. J. Greenwood for reading the manuscript and suggesting many improvements.

REFERENCES CITED

ALLER, L. H., 1961, The abundance of the elements: New York, Interscience Publishers.

AL'TSHULER, L. V., KRUPNIKOV, K. K., LEBEDEV, B. N., ZHUCHIKHIN, V. I., and BRAZHNIK, M. I., 1958, Dynamic compressibility and equation of state of iron under high pressure: Jour. Exptl. Theoret. Physics (U.S.S.R.), v. 34, p. 874–885. Trans. in Soviet Physics (JETP), v. 34, p. 606–614.

BASINSKI, Z. S., HUME-ROTHERY, W., and SUTTON, A. L., 1955, Lattice expansion of iron: Proc. Royal Soc., A, v. 229, p. 459–467.

BIRCH, F., 1939, The variation of seismic velocities within a simplified earth model in accordance with the theory of finite strain: Seismol. Soc. America Bull., v. 29, p. 463–479.

——— 1952, Elasticity and constitution of the earth's interior: Jour. Geophys. Research, v. 57, p. 227–286.

——— 1953, Uniformity of the earth's mantle: Geol. Soc. America Bull., v. 64, p. 601–602.

——— 1960, The velocity of compressional waves in rocks to 10 kilobars, pt. 1: Jour. Geophys. Research, v. 65, p. 1083–1102.

——— 1961*a*, Composition of the earth's mantle: Geophys. Jour., v. 4, p. 295–311.

——— 1961*b*, The velocity of compressional waves in rocks to 10 kilobars, pt. 2: Jour. Geophys. Research, v. 66, p. 2199–2224.

BOCKRIS, J. O'M., and RICHARDS, N. E., 1957, The compressibilities, free volumes, and equation of state for molten electrolytes; some alkali halides and nitrates: Proc. Royal Soc., A, v. 241, p. 44–66.

BOVENKERK, H. P., BUNDY, F. P., HALL, H. T., STRONG, H. M., and WENTORF, R. H., 1961, Preparation of diamond: Nature, v. 184, p. 1094–1098.

Boyd, F. R., and England, J. L., 1960, Aluminous enstatite: Carnegie Inst. Wash. Year Book 59, p. 49–52.

——— — ——— 1962, Effect of pressure on the melting of pyrope: *ibid.*, 61, p. 109–112.

——— — ——— 1963, Effect of pressure on the melting of diopside, $CaMgSi_2O_6$, and albite, $NaAlSi_3O_8$, in the range up to 50 kilobars: Jour. Geophys. Research, v. 68, p. 311–323.

Bridgman, P. W., 1948, Rough compressions of 177 substances to 40,000 kg/cm^2: Am. Acad. Arts & Sci. Proc., v. 76, p. 55–87.

——— 1949, Linear compressions to 30,000 kg/cm^2, including relatively incompressible substances: *ibid.*, v. 77, p. 187–234.

——— 1952, The physics of high pressure: London, G. Bell & Sons.

Bullard, E. C., 1957, The density within the earth: Kon. Nederland. Geol.-Minjb. Genoot. Verh. Geol. Ser., pt. 18, p. 23–41.

Bullen, K. E., 1936, The variation of density and the ellipticities of strata of equal density within the earth: Royal Astron. Soc. Monthly Notices, Geophys. Suppl., v. 3, p. 395–400.

——— 1940, The problem of the earth's density variation: Seismol. Soc. America Bull., v. 30, p. 235–250.

——— 1947, An introduction to the theory of seismology: Cambridge, Cambridge Univ. Press.

——— 1949, Compressibility-pressure hypothesis and the earth's interior: Royal Astron. Soc. Monthly Notices, Geophys. Suppl., v. 5, p. 335–368.

Bundy, F. P., 1959, Phase diagram of rubidium to 150,000 kg/cm^2 and 400° C.: Phys. Rev., v. 151, p. 274–277.

Chao, E. C. T., Fahey, J. J., Littler, J., and Milton, D. J., 1962, Stishovite, SiO_2, a very high pressure new mineral from Meteor Crater, Arizona: Jour. Geophys. Research, v. 67, p. 419–421.

Clark, S. P., 1958, Effect of pressure on the melting points of eight alkali halides: Jour. Chem. Physics, v. 31, p. 1526–1531.

——— 1959, Equations of state and polymorphism at high pressures, *in* Researches in geochemistry: New York, John Wiley & Sons.

———, Schairer, J. F., and de Neufville, J., 1962, Phase relations in the system $CaMgSi_2O_6$-$CaAl_2SiO_6$-SiO_2 at low and high pressure: Carnegie Inst. Wash. Year Book 61, p. 59–68.

Dachille, F., and Roy, R., 1960, High-pressure studies of the system Mg_2GeO_4-Mg_2SiO_4 with special reference to the olivine-spinel transition: Am. Jour. Sci., v. 258, p. 225–246.

Dorman, J., Ewing, M., and Oliver, J., 1960, Study of shear-velocity distribution in the upper mantle by mantle Rayleigh waves: Seismol. Soc. America Bull., v. 50, p. 87–115.

Edwards, G., and Urey, H. C., 1955, Determination of alkali metals in meteorites by a distillation process: Geochim. & Cosmochim. Acta, v. 7, p. 154–168.

Elsasser, W. M., 1950, The earth's interior and geomagnetism: Rev. Mod. Physics, v. 22, p. 1–35.

GILVARRY, J. J., 1956, Equation of the fusion curve: Phys. Rev., v. 102, p. 325–331.

——— 1957, Temperatures in the earth's interior: Jour. Atmos. Terr. Physics, v. 10, p. 84–95.

GUTENBERG, B., 1958, Wave velocities in the earth's core: Seismol. Soc. America Bull., v. 48, p. 301–314.

HIDE, R., 1956, The hydrodynamics of the earth's core, *in* Physics and chemistry of the earth, v. 1: New York, McGraw-Hill.

JEFFREYS, H., 1939*a*, The times of *P*, *S*, and *SKS*, and the velocities of *P* and *S*: Royal Astron. Soc. Monthly Notices, Geophys. Suppl., v. 4, p. 498–533.

——— 1939*b*, The times of the core waves: *ibid.*, p. 594–615.

——— 1959, The earth: 4th ed., Cambridge, Cambridge Univ. Press.

KAULA, W. M., 1961, A geoid and world geodetic system based on a combination of gravimetric, astrogeodetic and satellite data: Jour. Geophys. Research, v. 66, p. 1799–1812.

KELLEY, K. K., 1949, Contributions to the data on theoretical metallurgy. X. High-temperature heat-content, heat-capacity, and entropy data for inorganic compounds: U.S. Bur. Mines Bull. 476.

KING-HELE, D. G., 1961, The earth's gravitational potential, deduced from the orbits of artificial satellites: Geophys. Jour., v. 4, p. 3–16.

KNOPOFF, L., and MACDONALD, G. J. F., 1960, An equation of state of the core of the earth: Geophys. Jour., v. 3, p. 68–77.

LENNARD-JONES, J. E., and DEVONSHIRE, A. F., 1939, A theory of disorder in solids and liquids and the theory of melting: Proc. Royal Soc., A, v. 170, p. 464–484.

MACDONALD, G. J. F., 1962, On the internal constitution of the inner planets: Jour. Geophys. Research, v. 67, p. 2945–2974.

——— and KNOPOFF, L., 1958, The chemical composition of the outer core: Geophys. Jour., v. 1, p. 284–297.

——— and NESS, N. F., 1961, A study of the free oscillations of the earth: Jour. Geophys. Research, v. 66, p. 1865–1912.

MOREY, G. W., 1936, The phase rule and heterogeneous equilibrium, *in* Commentary on the scientific writings of J. Willard Gibbs: New Haven, Yale Univ. Press, p. 233–293.

NEWTON, R. C., JAYARAMAN, A., and KENNEDY, G. C., 1962, The fusion curves of the alkali metals up to 50 kilobars: Jour. Geophys. Research, v. 67, p. 2559–2566.

NOCKOLDS, S. R., 1954, Average chemical compositions of some igneous rocks: Geol. Soc. America Bull., v. 65, p. 1007–1032.

RAMSEY, W. H., 1948, On the constitution of the terrestrial planets: Royal Astron. Soc. Monthly Notices, v. 108, p. 406–413.

RICHTMEYER, F. K., and KENNARD, E. H., 1947, Introduction to modern physics: 4th ed., New York, McGraw-Hill.

RINGWOOD, A. E., 1959, On the chemical evolution and densities of the planets: Geochim. & Cosmochim. Acta, v. 15, p. 257–283.

——— 1962, Mineralogical constitution of the deep mantle: Jour. Geophys. Research, v. 67, p. 4005–4010.

Ross, C. S., Foster, M. D., and Myers, A. T., 1954, Origin of dunites and olivine-rich inclusions in basaltic rocks: Am. Mineralogist, v. 39, p. 693–737.

Rowlinson, J. S., and Curtiss, C. F., 1951, Lattice theories of the liquid state: Jour. Chem. Physics, v. 19, p. 1519–1529.

Sclar, C. B., Young, A. P., Carrison, L. C., and Schwartz, C. M., 1962, Synthesis and optical crystallography of stishovite, a very high pressure polymorph of SiO_2: Jour. Geophys. Research, v. 67, p. 4049–4054.

Segnit, R. E., and Kennedy, G. C., 1961, Reactions and melting relations in the system muscovite-quartz at high pressures: Am. Jour. Sci., v. 259, p. 280–287.

Slater, J. C., 1939, Introduction to chemical physics: New York, McGraw-Hill.

Stott, V. H., and Rendall, J. H., 1953, The density of molten iron: Jour. Iron and Steel Inst., v. 175, p. 374–378.

Strong, H. M., 1959, The experimental fusion curve of iron to 96,000 atmospheres: Jour. Geophys. Research, v. 64, p. 653–660.

Swanson, H. E., *et al.*, 1953, Standard X-ray diffraction powder patterns: Natl. Bur. Standards Circ. 539.

Tombs, N. C., and Rooksby, H. P., 1951, Structure transition and anti-ferromagnetism in magnetite: Acta Crystallographica, v. 4, p. 474–475.

Uffen, R. J., 1952, A method of estimating the melting-point gradient in the earth's mantle: Am. Geophys. Union Trans., v. 33, p. 893–896.

Urey, H. C., and Craig, H., 1953, The composition of stone meteorites and the origin of meteorites: Geochim. & Cosmochim. Acta, v. 4, p. 36–82.

Wackerle, J., 1962, Shock-wave compression of quartz: Jour. Appl. Physics, v. 33, p. 922–937.

Weir, C. E., 1956, Isothermal compressibilities of alkaline earth oxides at 21° C.: Natl. Bur. Standards Jour. Research, v. 56, p. 187–189.

Wentorf, R. H., 1962, Stishovite synthesis: Jour. Geophys. Research, v. 67, p. 3648.

Williams, A. F., 1932, The genesis of the diamond: v. 1, London, Ernest Benn Ltd.

Williamson, E. D., and Adams, L. H., 1923, Density distribution in the earth: Jour. Wash. Acad. Sci., v. 13, p. 413–428.

Willis, B. T. M., and Rooksby, H. P., 1953, Change of structure of ferrous oxide at low temperature: Acta Crystallographica, v. 6, p. 827–831.

Yoder, H. S., and Tilley, C. E., 1962, Origin of basalt magmas: An experimental study of natural and synthetic rock systems: Jour. Petrology, v. 3, p. 342–532.

JOHN A. O'KEEFE

Two Avenues from Astronomy to Geology

OUR CHAIRMAN for this symposium has set a difficult task for us, namely, to describe what sorts of investigations will be important in the future. He has specifically said that he is not willing to settle for a study of what is being done or for preliminary studies that outline what is just over the hill; he wants to know *why* these researches are going forward and what makes them profitable lines to pursue.

What makes his inquiry so difficult to answer is that, if we are doing really fundamental research, then the kind of usefulness that will come from our studies will depend on the results that we get. If we ask even a relatively simple question, with a "Yes" or "No" answer, there may be two quite different lines that future investigations will follow. With two fundamental questions we get four lines of investigation, and so on; with ten questions we have one thousand and twenty-four different lines, for each of which only five or ten words can be spared.

Hence, it appears to me that the reasonable way to respond to our chairman's invitation is to try to make a reasonable guess as to how the cat will jump at each of these points and to sketch the future on those lines. If we find that the prospects are good along this line, we may hope that they will be just as good when we have the real answers.

In what follows I shall try to present as well as I can the future in the field where astronomy leads into geology. The answers that I shall present are not chosen at random; they represent my best judgment. But, since the topics are quite controversial, you are welcome to consider them as no more than speculations serving to illustrate the value of further research.

The first avenue that I wish to discuss is the traditional one of the study of the earth's gravitational field. Along this road our predecessors

Dr. O'Keefe was born in Lynn, Massachusetts, and attended Harvard University and the University of Chicago, where he received the Ph.D. degree in astronomy in 1941. Subsequent work in geodesy, principally for the Army Map Service, led to his interest in the mechanics of the earth-moon system. Since 1958 he has been engaged in a physical and chemical study of the moon for the National Aeronautics and Space Administration.

found the sphericity of the earth, its flattening, the necessity of a massive core, and the phenomenon of isostasy, which plays such a great role in modern geophysical thought. In recent years, this road has continued to lead to useful ideas.

A few months after the launching of the first American satellites, in 1958, Diercks *et al.* (1958) announced the discovery that the flattening of the earth was significantly different from the value predicted on the basis of hydrostatic theory and the observed lunisolar precession. (See table 1.) At almost the same time Cornford (1958) in England and Buchar (1958) at Prague made the same discovery. The jump in the inverse oblateness that took place at that time, from values around

TABLE 1

INVERSE OBLATENESS OF THE EARTH, AS DETERMINED BY DIERCKS* *et al.* (1958)

SATELLITE	METHOD USED	
	Node	Perigee
1958 α..........	298.0 $\pm$0.3	297.8
1958 β_2.........	298.38$\pm$0.07	298.3

* Diercks was commanding officer of the Army Map Service at that time; he took an active interest in this work and the publication of these figures was an act of particular courage on his part, since they flew in the face of the accepted ideas both as far as geophysics is concerned and as far as military geodesy is concerned. His name appears on the Harvard Announcement Card only as commanding officer, Army Map Service, which causes some confusion in bibliographical reference.

297.3, which had been fashionable up to 1958, to the current values, near 298.3, has not been reversed in the ensuing four years; in fact, one can truly say that the work of the last four years on the determination of the flattening of the earth outweighs the work of the preceding four centuries by at least a factor of 100.

Toward the end of the same year, a group at NASA (O'Keefe and Eckels, 1958; O'Keefe, Eckels, and Squires, 1959) showed from some perturbations of the perigee height of Vanguard I that the earth is slightly asymmetrical in shape, the Southern Hemisphere being more flattened than the Northern. The result was confirmed and refined by Kozai (1959). Later, King-Hele and his co-workers (1961) showed that, as compared with an ellipsoid of revolution, the earth has a broad depression in the middle latitudes of both hemispheres.

The orderly way to discuss these peculiarities of shape is in terms of zonal harmonics. These are a set of ideal shapes, all rotationally symmetrical, to which any actual figure can be compared. Any rotationally symmetrical figure can be analyzed into a sequence of zonal harmonics by a process very similar to the analysis of a steady noise into a set of musical notes. Both processes are called harmonic analysis, and from the mathematical point of view the analogy is quite fundamental.

The surface that we analyze is not the topographic surface of the earth. It is, instead, the sea-level surface, which geodesists are accustomed to call the *geoid*. At sea it is mean sea level, and on land it is best defined as the prolongation of mean sea level as found by the techniques of spirit leveling. The significance of the geoid lies in the fact that it is an equipotential of the gravity field of the earth—including the earth's centrifugal force with gravity, as is customary in geodesy. Every property of the external gravity field then corresponds to some kind of a wiggle or hump in the geoid, and this is a great help in visualizing this surface (fig. 1).

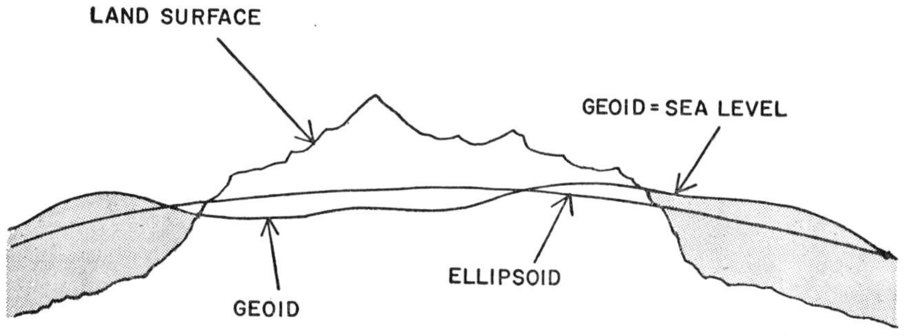

FIG. 1.—Geoid, ellipsoid, and topographic surface of the earth

The zonal harmonic of zero degree corresponds to the over-all size of the geoid; that of the first degree determines the position of the center-point in the axial, or north-south, direction. The second-degree harmonic corresponds to the tendency toward the shape of an ellipsoid of revolution. The third-degree harmonic corresponds to the simplest kind of north-south asymmetry in shape, namely, a tendency toward a pear shape, and the fourth harmonic to the formation of a depression (or, with the opposite sign, a ridge) in middle latitudes. More complicated harmonics of higher degrees correspond to more waves in the surface, at the rate of half a wave per degree of the harmonic. All the harmonics we have talked about so far follow parallels of latitude. Of the higher harmonics, King-Hele (1961) early found strong evidence of a measurable sixth harmonic, and Kozai (1960), at the Smithsonian, measured the fifth, and even the seventh and ninth (fig. 2).

In addition to the zonal harmonics, there is a much more numerous class of harmonics called the *tesseral* harmonics, which involve both latitude and longitude. Associated with each zonal harmonic of degree *n*, there are 2*n* tesseral harmonics. The most famous of the tesseral harmonics are the pair of the second degree that together determine how much the equator is out of round and in what longitudes the equatorial radius is greatest, that is, the effects usually discussed by saying that the earth has three unequal axes. It is much more difficult to determine the tesseral harmonics than the zonal because the rotation of the earth tends to blur their effect on a satellite. Recently, Kaula (1962) and Kozai (1962) have determined a large number of tesseral harmonics that confirm each other in the important question of size and the progression of size with the degree of the harmonic.

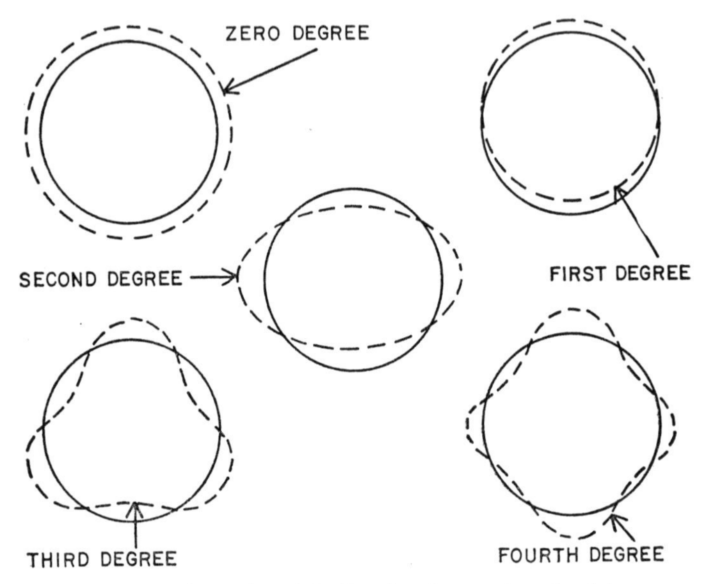

FIG. 2.—Harmonics of the first few degrees (zonal only)

The significant thing about these harmonics in the gravity field, and the corresponding waves of the geoid, is that the measured values do not agree with the values calculated on the theory of a plastic earth. The harmonic of zero degree is arbitrary, and that of the first degree depends on the choice of the origin of coordinates; therefore, nothing can be obtained from these two. But the second and fourth harmonics can be predicted, as mentioned, from hydrostatic theory and the measured values of the precession of the equinoxes. The same theory predicts zero values for the third and fifth harmonics and all higher harmonics. Prior to 1958 there had been no certain indication of measurable discrepancies between theory and observation, although it had long been clear that for very high harmonics the actual values were far from zero. Sir Harold Jeffreys (1952) had deduced by a careful analysis of the gravity data the probability of discrepancies of just about the size eventually found, but he was a voice crying in the wilderness.

Since these values are not predicted by hydrostatic theory, it is evident that the earth is not, as a whole, in hydrostatic equilibrium. The logical alternatives are either that it is in some kind of hydrodynamic motion or that it is kept in static equilibrium by forces resulting from mechanical strength, like an ordinary solid.

With regard to the possibility of hydrodynamic motion, the most completely developed theory is that of Vening Meinesz, described in Heiskanen and Vening Meinesz (1958, p. 369). From a consideration of the uplift of the Fennoscandian area after the last glaciation, these authors deduced a value of 10^{22} poises for the effective viscosity of the mantle in responding to stresses, and they gave an important formula for the time of relaxation of the earth after the removal of stresses:

$$t_r L = 6.3 ,$$

where t_r is the time of relaxation of the stresses, in thousands of years, and L is the width in megameters of a zone of stress, which is supposed to be infinite in length. If the length and the width of the zone of stress are equal, the constant becomes 9.9. For a second harmonic, the raised zone has a width of about 8 megameters and a length somewhat less than 40 megameters, so the time should be near 1,000 years. On the other hand, the height changes in the solid earth required to reduce the second harmonic to its equilibrium value amount to hundreds of meters (Jeffreys, 1952). Since it is perfectly clear that changes of sea level amounting to tens of meters per century are not occurring world-wide on any such pattern as this, we can be sure that some forces in addition to those that might be caused by viscous (10^{22} poises) flow are at work in the earth.

Two explanations offer themselves. Since we cannot account for the stresses by hydrostatic theory, we may choose between an explanation in terms of hydrodynamics and one in terms of the statics of solid bodies.

We begin by looking at the hydrodynamical possibility. Conceivably, the low harmonics of the earth's field may be sustained by viscous forces arising from movements within the earth. In this case, the stresses and their accompanying distortions of the earth's field will be permanent as long as the motions continue. The movements, being made against viscous forces, perform work. It is therefore logical to suppose that they take the form of convection currents, since this is the only way that has been suggested for converting the heat energy of the earth's interior into mechanical work. We are thus led to inquire whether the observed deformations of the gravitational field can be manifestations of convection currents.

We can use the known amplitudes of the harmonics to make a test of the convection theory. It happens that convection currents, like undulations of the geoid, can be analyzed into spherical harmonics. The convection current corresponding to each spherical harmonic is expected to exist more or less independently of the others, although, since there is only a finite amount of heat to be transported, the existence of one convection current makes it more difficult for others to come into being.

Considerable thought has been given by the supporters of the convection-current idea to the question of which harmonic is likely to be the strongest. There has been remarkable unanimity (Heiskanen and Vening Meinesz, 1958, chap. 11; Chandrasekhar, 1961, p. 266; Jeffreys, 1952, p. 331) in the conclusion that harmonics of the third degree are expected to predominate over all others. Those of the fourth degree are expected to be next, then those of the second degree, and higher degrees come tagging along behind.

The reason for this preference of the convection currents for third and fourth harmonics is not difficult to understand. The thickness of the mantle corresponds roughly to the size of a convection current (rising or falling) of the third or fourth degree. Such systems have the advantage over broader systems of the same depth that there is more vertical motion in a single rotation, in comparison with the horizontal motion. Since the horizontal motion involves friction but does not supply energy, it is a drag on the process. On the other hand, very narrow cells increase the vertical friction, since viscous forces depend on the *gradient* of the velocity. From mathematical reasoning, as well as common experience, it appears that the most efficient cells are approximately equal in all three dimensions. Hence, in the mantle we expect that convection will start with harmonics of the third or the fourth degree if the convection extends through the whole shell, and of higher degree if it is confined to some portion of the shell (fig. 3).

In table 2 we see Kaula's present estimates of the coefficients of harmonics of the earth's gravitational field, expressed in milligals of gravity anomaly at the surface of the earth. In order to make the comparison of various harmonics fairer, Kaula uses normalized harmonics, that is, harmonics that are so defined that the integral of the square of the surface harmonic over the surface of the earth is 4π when the coefficient of the harmonic is 1.

Two points emerge from this examination. The first is the presence of significant values for harmonics all the way from the second through the sixth degree. This is somewhat unfavorable to the convection explanation, since it would be expected on this basis that currents would appear to belong to some particular system that is presumably the most efficient. These would then so reduce the thermal gradient that no other system of currents could appear. It is a fact that, when convection currents are actually observed, they are found to form some sort of organized array, with a definite, characteristic size of the cell, at least when convection is beginning. Later, when the convective movement becomes more active, there may be a disorderly movement, but at first only the most efficient current systems can survive. Hence it is surprising that such an array of harmonics is observable. It is true that harmonics of degree n will produce stresses that depend not only on the

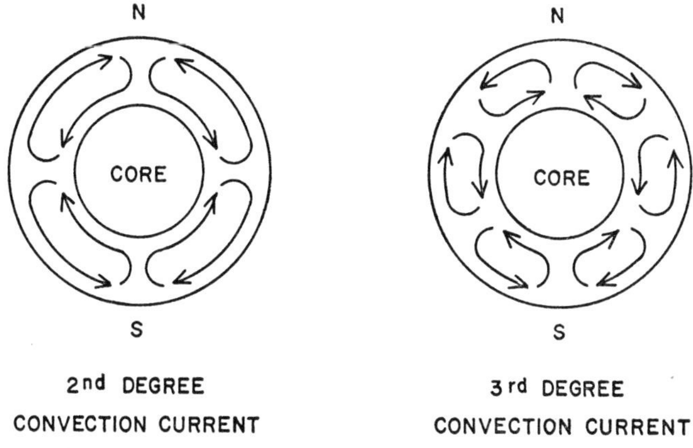

FIG. 3.—Convection currents in the mantle of the earth

harmonic itself but also on its derivatives. It turns out that the derivatives with respect to position on the sphere are of degree $n + 1$, so we would expect harmonics of two consecutive degrees, but hardly four. Notice that this argument did not trouble Vening Meinesz or Chandrasekhar; they could properly assume that the distribution of the continents was the result of a sequence of conditions in some of which the convection was of one degree, and in others of another degree. They could also invoke doubts about the stability of the poles or the uniqueness of convection currents to explain their discrepancies. These escapes are not open in the present case. We ought to be dealing with a single, more or less orderly, set of convection currents, and we very clearly are not.

A second difficulty is the fact that the largest single deviation from equilibrium is the value of J_2, the coefficient of the second-degree zonal harmonic. It is this coefficient that gives the preponderant size of the group of second-degree harmonics. The maintenance of this coefficient at its observed value would require a convection current that would rise at the equator and sink at the poles. It would be very considerably wider than deep. It is hard to understand how such a current could manage to survive competition with the third harmonics. To emphasize the point, the last column in table 2 shows reciprocals of Chandrasekhar's coefficient C_l, which measures the probability of the formation of convection currents of degree n. It is clear that both fourth- and, especially, third-degree currents are favored over second-degree currents.

On the whole, it appears most likely that the stresses in the earth's interior are not produced by the movement of the mantle material but are ordinary mechanical stresses that have not disappeared because the rock has a non-zero ultimate strength and can resist the stresses indefinitely.

TABLE 2

COMPARISON OF KAULA'S ESTIMATE OF COEFFICIENTS OF HARMONICS OF EARTH'S GRAVITY FIELD WITH C_l, PROBABILITY OF FORMATION OF CONVECTION CURRENTS OF DEGREE n, ACCORDING TO CHANDRASEKHAR

Degree of Harmonic	Sum of Squares of Normalized Coefficients of Potential, in Units of 10^{-12} (1)	Sum of Squares of Normalized Coefficients in Gravity, Unit Arbitrary (2)	Reciprocal of C_l for a Core Occupying Half the Radius (3)
2	33.2	33.2	4.585
3	4.9	19.6	5.198
4	1.3	11.7	4.660
5	(0.28)	(4.5)	3.741
6	(0.26)	(6.5)	2.864

In calculating the second-degree coefficient of column 2, the value of the zonal harmonic was taken as the difference between the equilibrium value ($J_2 = -0.0017707$) and the actual value ($J_2 = -0.0010823$, as obtained by Kaula). No allowance was made for the small equilibrium value of the fourth zonal harmonic.

Kaula gives only 3 out of 11 fifth-degree coefficients and only 9 out of 13 sixth-degree coefficients; the missing values were assumed to have the same average value as the measured coefficients of the same degree.

Column 3 is obtained by multiplying column 2 by $(n - 1)^2$; see Jeffreys (1952, p. 133).

In this case, we must ask ourselves how the stresses arose in the first place. In particular, what produced the observed load on the equatorial belt (over and above the equilibrium value of the bulge) and the observed lightness in the polar regions? Several explanations may be offered. Conceivably, the stresses might arise from the melting of the ice-age snows. Again, they might be due simply to the fact that throughout its history the earth has been warmer at the equator than at the poles, and the interior is therefore denser in these regions. The increase in density, resulting from climate variation, extends surprisingly far into the earth, according to Jeffreys (1952, p. 326). Again, the stresses might be a consequence of the famous von Zeipel theorem (Eddington, 1930), which says that a rotating body in which heat is flowing cannot be in hydrostatic equilibrium unless energy is being produced in the interior in a particular and highly implausible way. This astonishing theorem depends on the fact that in hydrostatic equilibrium layers of equal pressure and density are expected to follow the gravitational potential, so that equipotential surfaces are likewise surfaces of constant pressure and density. Since, in any substance possessing an equation of state, the temperature is a function of pressure and density, it is clear that in hydrostatic equilibrium the temperature must also follow these equipotential surfaces. But if these conditions obtain initially, they cannot persist. The equipotential surfaces are farther apart at the equator than at the poles; hence the flow of heat will be more rapid at the poles, and the cooling will go more rapidly there. The two sets of equations—hydrostatic and heat-flow—cannot be simultaneously satisfied. The above-mentioned implausible distribution of sources provides only a formal escape from the dilemma; there is no reason to believe that the actual distribution has this form.

In the case of the actual earth, von Zeipel's theorem seems to work out as follows: If we begin with an earth that is fluid, and allow it to cool, then, as long as the currents can flow, they will bring us close to the hydrostatic distribution. But, as soon as the currents stop, the polar regions will begin to cool faster than the equatorial regions. This will result in an earth that is more flattened than it ought to be for the given polar moment of inertia.

A fourth explanation undoubtedly furnishes at least a part of the observed extra flattening. The retardation of the earth's velocity of rotation has been observed; it implies that throughout geological time the earth must have been getting less and less flattened, since it rotated more and more slowly. MacDonald has estimated (personal communication; cf. Munk and MacDonald, 1960, p. 280) that the present flattening is that appropriate to the velocity of rotation that the earth had some fifty million years ago.

We see that it is very difficult to isolate and evaluate any causes of the deformation of the earth's figure so long as we are confined to those deformations (the zonal harmonics) which depend on latitude alone. We shall undoubtedly understand these problems better when we have improved numerical values for the longitude-dependent (tesseral) harmonics.

I cannot claim to have satisfied your chairman's request for a definition of problems by pointing out the need for more accurate values of the tesseral harmonics. The very astute and patient celestial mechanicians who are now working on these problems are fully conscious of the need. They are bedeviled at present by the difficulty of separating the effects of tesseral harmonics from those of datum errors at the tracking stations and of disentangling the tesseral harmonics from one another. These difficulties do not seem insuperable, however. It is my belief that the tracking of manned satellites, which has been uniquely thorough and complete, will soon permit the identification and measurement of datum differences with all the precision needed. With this help, and possibly with the help of the kind of continuous tracking that is expected on EGO, it may soon be possible to go further. What we can see plainly, however, is that at some past time stress differences were introduced into the deep interior of the mantle, and these stresses have not yet relaxed.

From here on, the avenue is blocked by trees and underbrush. To get an idea of what may be beyond, we go back and start over again along what seems to be an entirely unrelated path.

The second avenue from astronomy to geology is the study of the geology of the moon—of selenology. Remote as this may seem from the contemporary concerns of the geologist, it will undoubtedly be so prominent a part of geology ten years from now that it seems to me that every thoughtful young student of geology must already be asking himself whether he is going into terrestrial or lunar geology. Along this way will be found, I believe, a respectable fraction of the great discoveries in geology during the next twenty years.

Along the first part of the road, the study of craters of the moon, it has recently become much clearer where we are going. The discovery of coesite in Meteor Crater by Chao and Shoemaker (1960) has given us a means of identifying meteor craters everywhere. An immediate result was the identification as an impact feature of the Ries Crater, which is located in central Germany and is twenty times the size of Meteor Crater—settling in the process a decades-long debate in the German geological literature (Shoemaker and Chao, 1961). Another result was the explanation of the appearance of diamonds in some of the meteorites from Meteor Crater (Anders, 1961), as having apparently resulted from the enormous pressures produced by shock. With this explanation there toppled to the ground the elaborate theoretical structure erected by H. C. Urey (1956), largely on the basis of the presence of diamonds in certain meteorites, in which bodies of lunar size were built up, broken down, and rebuilt. Since we have found and are finding huge impact craters on the earth, it is no longer possible to doubt that at least some of the huge craters that we see on the moon are of impact origin.

We have not, however, as yet identified on the earth any craters of the size of Copernicus, which is 90 kilometers in diameter, much less, features like Clavius, which is over 200 kilometers across. From the abundance of large craters on the moon it appears possible that over half the surface of the earth was at one time or another within the walls of one of the great craters. Rocks resulting from such impacts have undoubtedly been picked up, studied, and named, like the suevite of the Ries. The possibility that especially puzzling geological formations are of impact origin is one that must hereafter be kept in mind.

Beyond this point the road is not clear. Are the lunar maria lava flows, ash flows, or pools of dust? Does isostasy exist on the moon? Is its interior hot? Are the small domes laccoliths, or ice domes, or basaltic shield volcanos? Above all, at what point in the history of the solar system does the history of the matter that now forms the moon branch off from the history of the earth?

I shall try to show how these questions could be answered if it were true that certain small stones called tektites found in Texas and other places actually did come from the moon. I shall not attempt here to argue the question whether this hypothesis is correct, although I think that the aerodynamic evidence is compelling, but only to use the hypothesis as a guide in threading the way through the maze of possibilities. At the least, it will illustrate the enormous value to our problem that even a limited number of damaged samples from the moon can have.

In the first place, studies of the abundances of isotopes of lead, thorium, and uranium (Tilton, 1958), and especially of rubidium and strontium (Pinson and Schnetzler, 1961), have shown that tektites have ages of the order of a few hundred million years since they were differentiated. The work has been done carefully, in large laboratories, and by competent experimenters. The present abundance of rubidium 87 in tektites should have generated far more strontium 87 (over and above the initial amount) than is now observed if the rubidium had coexisted with the strontium for the 4.5-billion-year accepted age of the solar system, or even the 3-billion-year age of the oldest rocks. Hence, the material of the tektites must have been derived from some presumably more basic material a relatively short time ago, as cosmologists count time. If the moon was hot for the first nine-tenths of its history, it is probably still hot, and volcanic activity will be seen again, though not necessarily by us.

Again, because of the youthfulness of tektites, there must be some reasonably extensive youthful formations on the moon from which tektites could come. It is quite clear that the youngest extensive formations on the moon are the dark areas called the *maria;* they have about one-tenth as many craters as the brighter areas, called the *terrae.* On this ground alone, Kreiter (1960) has suggested that they are a few hundred million years old. There is a reasonably good fit between Kreiter's ages and the ages of the tektites, and we are led to conclude that tektites are samples of the maria.

If this idea is correct, then the abundance of impact craters on the earth must be very large, in fact surprisingly large. There are about 150 craters on the moon as large as or larger than Copernicus. The earth has 16 times the area of the moon and 30 times the area of its visible side. Hence, it should have something like 4,500 craters the size of Copernicus, or roughly one for every million years of its existence. Allowing for the probability that impacts in the sea would be lost, there is still one impact every 4 million years on land, at a scale such as to make a crater 90 kilometers across, or perhaps a little smaller if we allow for the effects of the earth's gravity.

This is enough to give pause to the boldest. I know of only one indication that this may be the proper order of magnitude, namely, the existence of the Ries Crater in central Europe, with an age of 15 million years. Considering the very small size of Europe, it appears to me highly suspicious that this area should have given us our only lunar-size crater, unless such craters are really quite common and have merely been recognized first in Europe. In this case, there is a great harvest waiting for some bold spirits.

If the maria are really the sources of tektites, they are very clearly not lava flows in the usual sense, since tektites are highly acid materials. Acid lavas rarely give rise to widespread lava flows, and, when they do, their high viscosity makes the flows very rough and irregular on the surface. Mare Crisium, on the other hand, is remarkably smooth, with average slopes of less than half a degree over hundreds of kilometers, and it is reasonably typical of lunar maria. It could hardly be an obsidian flow, especially since the decreased lunar gravity would make an obsidian flow even rougher than it is on the earth.

Neither could it be a dust pool if it is the source of tektites. A dust pool would have approximately the same isotopic constitution as the land from which it came, which is in this case the lunar terrae. If it is hard to swallow the idea that the maria are only a few hundred million years old, it is practically impossible to suppose that the terrae are so young; for, in that case, we must multiply the assumed rate of crater formation by a factor of 10.

The only remaining alternative seems to be that the maria are ash flows, or ignimbrites, as they are sometimes called. This idea fits reasonably well; there is even the curous fact that craters buried beneath the surfaces of the maria seem to show in a ghostly way at the top, just as buried drainage can be detected at the top of an ash flow. The flow was porous, and, when the porosity was squeezed out, there was more compression in the valleys than in the hills (Boyd, 1961).

In the same way, we are tempted to call the lunar domes laccoliths. I am referring here to the large domes, which are some 10 kilometers in diameter, with sides that slope at an angle of 2–5 degrees. Laccoliths would be expected to occur in an area of acid vulcanism. Moreover, the layered structure of the country rock that is implied by the notion of a laccolith might possibly be formed by the ash flows that we have imagined as having formed the maria. Domes have so far been found only in maria.

Similarly, one is led to interpret the system of narrow ridges that is seen on the shore of Mare Imbrium, following the grain of the country, as resulting from the outflow of an acid lava from a system of fault cracks. This explanation was suggested by Shaler (1903), who spent some thirty years studying the moon's surface. He specified that he thought they were trachytic.

In addition to this problem of the interpretation of lunar structure, the mere presence of large masses of acidic rock on the moon is of fundamental significance for our ideas of the origin of granite magma either on the earth or on the moon. If granite is somehow a metamorphosed sedimentary rock, it cannot exist on the moon, no matter what the cooking process. If tektites are really from the moon,[1] then it is hard to escape the conclusion that there are layers of granitic composition that are deep enough to fill the maria. Since the craters under the mare surfaces do not stick up except near the shores, the maria must be kilometers deep. If this is all granite, or granite-like rock, then clearly it is possible to produce great masses of granite magma without the intervention of any sedimentary processes whatever. This point has been made by Shoemaker (1962). It would, as I understand it, go a long way to support the ideas of the late N. L. Bowen (1928) and his successors Tuttle and Schairer.

At the end of our second avenue, we seem to see something else looming up before us. It has often been said that tektites cannot come from the moon because they are too much like the earth in their chemistry. The resemblance is certainly remarkable (Taylor and Sachs, 1960), but it is a resemblance of a peculiar kind. It is a resemblance more to

[1] The derivation of tektites from the maria presents some chemical problems. The chemistry of tektites differs in several respects from that of the acid products of "petrogeny's residua system" (Bowen, 1928): tektites contain about half as much K_2O and Na_2O and several times as much CaO and MgO as do typical acidic terrestrial rocks. Further, the SiO_2 content of some tektites is higher than that of any genuinely magmatic rock. The deficiency in alkalis and the excess of lime and magnesia might be explained by differential volatilization at some time during the history of the tektite (Cuttita *et al.*, 1962). The silica excess might be a more serious problem. Mechanical concentration of quartz phenocrysts in pyroclastic rocks has been reported and discussed (Pettijohn, 1957, p. 333); an analogous process (a sort of elutriation) might be operative during lunar ignimbrite extrusion. This process may be relatively efficient in the low gravity field of the moon, which is equivalent to decreasing the density of the fine component (cf. Leva, 1959, p. 118). Alternatively, the high silica content of some tektites could indicate their derivation from portions of lunar ignimbrites in which silica had been concentrated by a hydrothermal process. Because the physical parameters of the process by which tektites might have been derived from the moon are largely unknown, a discussion of the details of their chemistry does not seem appropriate at this time.

average terrestrial rocks than to particular rock samples. One hundred and seventy-five years have passed since the first attempts were made to find the mother-lode from which the tektites come, and all such attempts have failed. One of the most remarkable differences between terrestrial rocks and tektites seems to be the greater diversity of terrestrial rocks. When we reduce these diversities by taking averages, then, whether we take an average sandstone, with Urey (1959), or an average shale (diluted with quartz), with Taylor, Cherry, and Sachs (1960), or igneous rocks, with Cuttita and his co-workers (1962), we get close agreement. The agreement is particularly evident if we compare with the meteorites, which show us just how far off the crustal standard a rock can get. Tektites are remarkably like earthly materials, not only in their chemistry, but even in their detailed isotopic makeup.

We can hardly avoid the conclusion that if tektites are from the moon, then the moon is from the earth. The resemblances of trace-element abundances, of differentiation processes, and of age are too striking to be accidental. Every attempt to find a decisive chemical or nuclear difference between tektites and terrestrial rocks has failed.

It no longer appears impossible that the moon was derived from the earth by simple rotational breakup. If the moon were somehow put back into the earth, the angular velocity of the combined body would approach the velocity of breakup. However, it would not reach that velocity, and in former times that fact was regarded as fatal. But in recent years we have come to realize that stars and planets can, under some circumstances, lose angular momentum through magnetic interaction. There have even been some who have suggested that this is a principal cause of the slowing-down of the earth's rotation (Munk and MacDonald, 1960, p. 227). Without subscribing to that idea, we may speculate that, particularly in the early days of the solar system, the magnetic field of the sun, and likewise that of the earth, was stronger than now and that in some way a portion of the angular momentum of the earth-moon system got lost.

This leaves us free to imagine that the moon might have broken off from the earth as a simple result of the increased velocity of rotation that may have occurred when the core formed. The mechanics of the breakup of a liquid or plastic body under the influence of gravitational forces is a famous and incompletely solved problem of gravitational theory (Darwin, 1898). So far, almost all work has been done on the hypothesis of a homogeneous body. Lyttleton (1953) came to the conclusion that it was not possible for a liquid body to break into two parts as different in size as the earth and the moon and, further, that, if breakup occurred, the smaller portion would go off into space and never return. It would obviously be interesting to see whether the same conclusions hold in the case of an inhomogeneous body such as the earth.

Going back now to our first avenue, we thought we saw evidence at the end of the road of some large event that had stressed the earth to the bottom of the mantle. Is it possible that this event was in fact the birth of the moon? We would expect that this event would have been preceded by a marked extension of the earth to the form of an ovoid disc. Is it conceivable that in the slight triaxiality of the earth we have the vestiges of this great event?

Is it possible that the great circum-Pacific fault zone, which extends backward under the continents, represents the surface along which the great bulge sank back into the earth?

We have come to the end of our two avenues. We have seen the possibility that they lead us to the notion of a single great event. Let me say again that what is most important here is not the conclusion that may be suggested but the roads toward it.

In particular, I wish to draw your attention to the enormous importance for the future of geology of the investigations now going on in the high and remote field of celestial mechanics and, above all, to the decisive importance to geology of a returned lunar sample.

References Cited

Anders, E., 1961, Comments on the origin of natural diamonds: Astrophys. Jour., v. 134, p. 1006.

Bowen, N. L., 1928, The evolution of the igneous rocks: New York, Dover, repr., 1956.

Boyd, F. R., 1961, Rhyolite Plateau, Yellowstone Park, Wyoming: Geol. Soc. America Bull., v. 72, p. 387–426.

Buchar, E., 1960, Determination of the flattening of the earth by means of the displacement of the node of the second Soviet satellite (1958) : Presented at the July, 1958, CSAGI General Assembly, Moscow, IGY Annals, v. 12, p. 174.

Chandrasekhar, S., 1952, The onset of convection by thermal instability in spherical shells: Philos. Mag., ser. 7, v. 44, p. 233–241, correction, p. 1128, 1129.

Chao, E. C. T., Shoemaker, E. M., and Madsden, B. M., 1960, First natural occurrence of coesite: Science, v. 132, p. 220–222.

Cherry, R. D., Taylor, S. R., and Sachs, M., 1960, Major element relationships in tektites: Nature, v. 187, p. 680, 681.

Cornford, E. C., 1960, Comparison of orbital theory with observations made in the United Kingdom on the Russian satellites: Presented at the July, 1958, CSAGI General Assembly, Moscow, IGY Annals, v. 12, p. 151–176.

Cuttita, F., Carron, M., Fletcher, J., and Chao, E. C. T., 1962, Chemical composition of bediasites and philippinites: prepr., February, 1962.

Darwin, G. H., 1898, The Tides: San Francisco, reprinted by W. H. Freeman & Co., 1962.

DIERCKS, F. O., O'KEEFE, J. A., HERTZ, H. G., and MARCHANT, M., 1958, Oblateness of the earth by artificial satellites: Harvard Coll. Observ. Announcement Card 1408, June 24, 1958. See also IGY Annals, v. 12, p. 176.

EDDINGTON, A. S., 1930, The internal constitution of the stars: Cambridge, Cambridge Univ. Press.

HEISKANEN, W. A., and VENING MEINESZ, F. A., 1958, The earth and its gravity field: New York, McGraw-Hill.

JEFFREYS, H., 1952, The Earth: 3d ed., Cambridge, Cambridge Univ. Press.

KING-HELE, D. G., 1961, The earth's gravitational potential, deduced from the orbits of artificial satellites: Royal Astron. Soc. Geophys. Jour., v. 4, p. 3–16.

KOZAI, Y., 1959, The earth's gravitational potential derived from the motion of satellite 1958 β_2: Smithsonian Astrophys. Obs. Spec. Rept. 22.

——— 1962, Numerical results from orbits: Smithsonian Astrophys. Obs. Spec. Rept. 101.

KREITER, T. J., 1960, Dating lunar surface features by using crater frequencies: Astron. Soc. Pacific Pub., v. 72, p. 393–398.

KUIPER, G. P., 1959, The exploration of the moon, p. 273–313 *in* Vistas in Astronautics, v. 2: London, Pergamon Press.

LEVA, M., 1959, Fluidization: New York, McGraw-Hill.

LYTTLETON, R. A., 1953, The stability of rotating liquid masses: Cambridge, Cambridge Univ. Press.

MUNK, W. F., and MACDONALD, G. J. F., 1960, The rotation of the earth: Cambridge, Cambridge Univ. Press.

O'KEEFE, J. A., and CAMERON, W. S., 1962, Evidence from the moon's surface features for the production of lunar granites: Icarus, v. 1, p. 271–285.

——— and ECKELS, A., 1958, Perturbation in the eccentricity of 1958: Harvard Coll. Observ. Announcement Card 1420.

———, ———, and SQUIRES, R. K., 1959, Pear-shaped component of the geoid from the motion of Vanguard I: Science, v. 129, p. 565, 566.

PETTIJOHN, F. S., 1957, Sedimentary Rocks: New York, Harper & Bros.

PINSON, W. H., and SCHNETZLER, C. C., 1961, Rb-Sr correlation studies of tektites: Jour. Geophys. Research, v. 66, p. 2553.

SHALER, N. S., 1903, A comparison of the features of the earth and the moon: Smithsonian Contr. Knowledge, v. 34, separately printed by the Smithsonian Institution, Washington, D.C.

SHOEMAKER, E. M., 1962, Exploration of the moon's surface: Am. Scientist, v. 50, p. 99–128.

——— and CHAO, E. C. T., 1961, New evidence for the impact origin of the Ries basin, Bavaria, Germany: Jour. Geophys. Research, v. 66, p. 3371–3378.

TAYLOR, S. R., and SACHS, M., 1960, Trace elements in australites: Nature, v. 188, p. 387–388.

TILTON, G. R., 1958, Isotopic composition of lead from tektites: Geochim. & Cosmochim. Acta, v. 14, p. 323–330.

UREY, H. C., 1956, Diamonds, meteorites and the origin of the solar system: Astrophys. Jour., v. 124, p. 623–637.

——— 1959, Chemical composition of tektites: Nature, v. 183, p. 1114.

W. S. FYFE

Experiment and the Crust of the Earth: Problems and Approaches

I WAS INDEED HONORED to be asked to contribute to this symposium. But, having accepted, I was really a little frightened to find, at a later date, the task assigned to speakers. We have been asked to consider new, exciting, and important questions. I do not feel like a geochemical seer and can only hope that, even if I do some predicting, few will agree with me. We live in an age of explosive scientific expansion. Experiments and theories have ceased to be significant, fundamental, or critical—all have become fabulous or exciting, and many constitute break-throughs. Language changes, but perhaps part of this change stems from the hope of personal identification in the mighty machine of twentieth-century science.

Fortunately, we have been allowed to consider the philosophy of our science, and perhaps I might start along these lines. The number of really important advances in science, as opposed to technology, tends to be rather limited. If we were asked what have been the most exciting scientific advances of this century, I think most of us would think of quantum theory and relativity. But how did such new concepts develop? The birth of quantum theory is normally associated with Planck and a publication in *Annalen der Physik* in 1901. But the problem on which Planck was working was far from new. Measurements on black-body radiation were well started by 1850. Kirchoff's law was formulated in 1860, the Stefan-Boltzmann law in 1879, Wein's law in 1893, and the Boltzmann distribution law in 1886. Thus we can assume a fifty- or, at least, twenty-year induction period leading to Planck's momentous advance. But for our purposes this was only the start. Bohr applied the

Dr. Fyfe was born in New Zealand and educated in the University of Otago, where he received the Ph.D. degree in chemistry in 1952. He has been an associate professor in the Geology Department of the University of California at Berkeley since 1959. His principal contributions to the earth sciences have been in the application of chemical principles to problems of metamorphic petrology.

theory to atoms in 1913. The mathematics of atomic-molecular quantum systems came with the work of Dirac, De Broglie, Heisenberg, and Schrödinger around 1926. And the first accurate solution of a real system of interest, the hydrogen molecule, was published by James and Coolidge in 1937, in the first volume of the *Journal of Chemical Physics*. Perhaps the full impact of all this became generally significant in chemistry and geochemistry only with the publication of Pauling's *Nature of the Chemical Bond* in 1940. I think that this example does show us that the development of something new often arises from old problems and that the recognition, application, and infiltration of the new idea through all science may be a most sluggish process.

Often, too, the significance of the insignificant observation may be lost for many years. Would DDT have been discovered as a potent insecticide, whose use has saved millions of lives, if it had not been synthesized by Zeidler in 1874? Such a question cannot be answered, but if Mueller, who gained the Nobel prize for the discovery of DDT in 1948, worked like many of us, he had a fair idea of the type of compound required and found it. Thus, in searching for the test of some idea, we must often use the dullest data. Were it not for this most unscientific lack of prediction, built into the very certainty of scientific uncertainty, our lot would be a dull one. So I do not wish to predict too much in this lecture, for I am sure I would waste your time. Science, perhaps even more than feminine whims, has fashions, but it also tends to return to the old ones. Therefore, I shall consider mainly old problems, old crustal problems, still nagging, which are by-passed for the moment by the moon and moholes but which must be solved. If history is with me, the old problems are perhaps still the most exciting.

I shall start by briefly reviewing some ideas that I consider to have been fundamental in bringing us to our present situation. Any geochemist must first pay tribute to all who have developed the techniques of one of the most demanding exercises in analytical chemistry: rock and mineral analysis. In 1900 Hillebrand published "United States Geological Survey Bulletin No. 176," containing one of the first accounts of complete silicate analysis. Hillebrand stressed the need for complete analysis, a rarity up to that time. Clarke and Washington's "U.S.G.S. Professional Paper No. 127," 1924, indicates the mass of data accumulated over a rather brief period—a period that, as Hillebrand remarks, was organic chemistry's heyday, when inorganic analysis was almost despised.

In the 1920's came the introduction of X-ray and optical spectrography, the latter being brought into fruition by the school of V. M. Goldschmidt at Göttingen. Trace-element analysis now became possible on a routine basis, and the second stage of inorganic geochemistry had arrived. Since that time, non-destructive and instrumental techniques have gained in significance, with further development of the X-ray spectrograph, flame photometer, and neutron activation. Today, the development of the electron microprobe has brought us to the third stage of inorganic analysis. I shall return to this later.

Geology is in large part a science of solids, and understanding of the solid state owes much to our science—but it should owe much more. After Laue's discovery of X-ray diffraction, and the fuller interpretation by the Braggs (1912–1913), it was again the genius of Goldschmidt, who appreciated the geochemical significance of crystal chemistry, that led to the development of these ideas in the earth sciences and, in particular, the importance of ionic radii. If one looks at F. W. Clarke's heroic attempt at silicate-structure analysis, published as "U.S.G.S. Bulletin No. 588," ironically in the year 1914, it is apparent

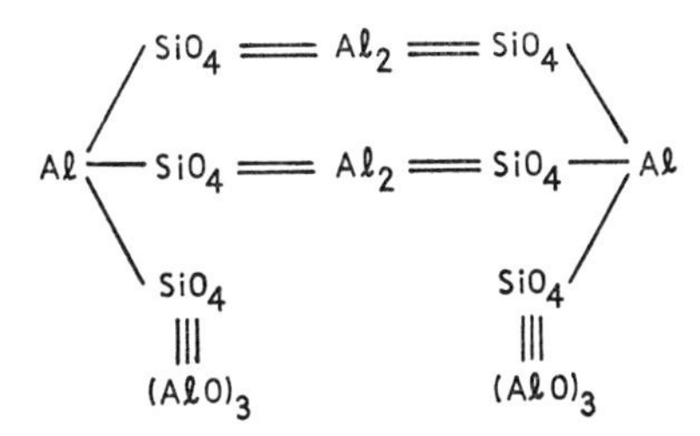

Fig. 1.—An "organic" structural formula for kyanite. (After F. W. Clarke, 1914.)

that X-ray crystallography saved the student of mineralogy a great deal of meaningless effort (see fig. 1). The publication of Bragg's *Atomic Structure of Minerals* in 1937 marks the real introduction of crystal structures to students of the earth sciences.

Other studies in the solid state, such as those involving lattice imperfections, have important geological consequences. Perhaps the recognition of many common types of defect should have originated in our science. Blue fluorites, colored calcites, etc., contain the clue to *F* centers and other defects. Mott and Gurney's *Electronic Processes in Ionic Crystals*, published in 1948, gives a clear account of the development of this field dating from about 1920. Solid reactivity is so intimately connected with defects that it is still perhaps puzzling why so little of this subject is discussed in our present courses. Any many extravagant claims regarding the powers of solid diffusion would not have been made if the available data had been critically examined. Crystal growth has also become a more intelligible process since the recognition of the significance of dislocations through the work of Frank in 1952 and of others.

In a vast number of geological processes we are concerned with the energy of lattice formation and reorganization. Theories of lattice energy were well advanced by the 1920's with the work of Born, Madelung, and Meyer, but how little attention the earth scientist has given this subject! I am convinced that we have an enormous contribution to make to the understanding of binding forces in solids. I shall return to this later.

A great deal of our analysis of geological processes assumes that, in general, a close approach to local equilibrium is achieved when mineral phases form. If this assumption is largely true (it is most certainly not generally true), then clearly the methods of classical thermodynamics must guide in all studies and analyses. But, while much lip service is paid, the full value of such analysis has not yet been achieved. To many, the phase rule is the beginning and end of thermodynamics in the geological context, but, as Denbigh (1955) states, "greater physical insight into a particular problem may often be obtained by making direct application, not of the phase rule itself, but of the conditions of equilibrium on which it is based." In this field we again see the slowness of the infiltration of a "break-through."

Classical thermodynamics was well advanced by the middle of the last century through the work of Joule, Clusius, Kelvin, and others. Willard Gibb's famous treatise appeared in 1876, but it was the industrial chemists of Germany, and in particular Haber, who convinced chemists that this was more than an academic exercise. It is possibly true that chemists paid little routine attention to this subject before the contribution of G. N. Lewis and the Berkeley school. In our science it was again Goldschmidt who, in 1911 in his famous doctoral thesis, developed the phase rule in its geological context, but it was the work of the Carnegie Institution, brought to fruition in Bowen's *Evolution of the Igneous Rocks* in 1928, that forced the petrologist to appreciate the value of this approach. Certainly in my own student days, which were not so long ago, the use of phase diagrams and the phase rule was something very modern—in fact only sixty years old.

When we consider modern geochemical approaches to sedimentary phenomena, the same story appears again. Modern electrolyte theory began in large part through the studies of Arrhenius and Ostwald around 1890. In due course these led to the development of the theoretical treatment of ion activities of Debye and Huckel in 1923. These theories, largely complete by this time, are capable of rationalizing many of the problems regarding precipitation and solution of sparingly soluble salts (minerals). The study of the significance of the electromotive force of chemical cells dates back to Gibbs in 1875 and Von Helmholtz in 1882, who proposed the fundamental relation between electromotive force (E) and free-energy change (ΔG). With the relation, $\Delta G = -nFE$

(where n is number of equivalents per mole, and F is the Faraday), aqueous oxidation-reduction chemistry was placed on a firm basis. Then follows the long period of data collection guided by the Lewis-Latimer school, which combined all aspects of solubility, oxidation-reduction, and complexing equilibria into a framework that allowed the simplification and correlation of the maze of data involved in aqueous inorganic chemistry. Although there were minor incursions previously, it is Garrels of Harvard who has demonstrated the value of these approaches. Again we see a lapse of fifty years between theory and data collection and a lapse of thirty years for geological appreciation.

This review could be continued much further. Organic geochemistry depends in no small measure on the techniques of chromatography, and in this field chemists also were slow to recognize the importance of the work of the Russian botanist Tswett, whose amazingly detailed studies were published in 1906. The foundations of isotope geology, with its many aspects, are also to be found in the early part of this century.

Perhaps this survey of the development of some features of modern geochemistry may lead us to doubt a little how successful any attempts at prediction of future trends may be. General themes tend to be developed, accepted, and applied quite slowly in science. Few were appreciated by many at the time of their birth. Geochemistry today has borrowed heavily in the past from the fundamental physical sciences, and perhaps it is to these that we should look for the next major advances. There are thus some difficulties in the task proposed by those who conceived the pattern of this symposium.

In proceeding, I wish to consider some obvious gaps in our knowledge: fundamental gaps the filling of which will require the integration of theories and techniques of all the physical sciences. Often, we may require not new theories but extension of existing theories. To make the discussion a little less illogical, I will chop up our subject into some arbitrary divisions.

One significant fact must be borne in mind when we approach the crust of the earth today. Within the crust, the normal pressure range is of the order 20,000 atmospheres, and the temperature ranges up to 1,400° C. This range is now within reach of apparatus that can be modified to carry out many of the operations of conventional 25° C. physical chemistry. We cannot be completely satisfied with this, for the origin of some crustal rocks may reflect deeper environments. However, some observations have been made under the extreme conditions of 3,000° C. and 200,000 atm., which must satisfy almost all our crustal needs. Certainly there are instrumental difficulties, but these will be solved if a convincing need arises; the choice of the significant problems and the crucial experiments are now the most important.

Chemical Analysis and Element Distribution

Whenever the concentration of a minor element in a host rock or mineral has been determined previously, its exact mode of disposition in this host has been more or less indeterminate, except in rare cases. Thus the common generalizations regarding trace-element distribution have assumed a homogeneous solution in a given crystal lattice. On the basis of this assumption and of Goldschmidt's study (1937) of ionic radii, a number of basic rules governing distribution and fractionation have been suggested, and these rather frequently appear to offer a good explanation of observations. Goldschmidt's views have been modified in minor ways largely by considering departures from his essentially ionic model of a crystal. There have been some observations, particularly with radioactive elements, indicating that homogeneous distribution in the host may not always be the case, and uranium leaching studies, for example, indicate that this element is concentrated on grain boundaries, etc.

Today, in this field of study, we have arrived at the third major stage. Few geologists are now unaware of the potentialities of the electron microprobe, developed in France by Castaing and in part introduced into petrology by the Cambridge school of mineralogy and petrology. The simple fact that electrons may be focused in a manner impossible with photons allows us to study very small areas of finite thickness. In fact, a chemical analysis is now possible on a volume of material of the order of 10^{-15} cm.^{3}. From now on, if we are to progress, a trace element study without a complimentary probe study will be of restricted value. If the trace element is concentrated in discrete grains of submicron size, this can be ascertained. But the use of electrons in some future problems may go beyond the present-day probes. The electron microscope may provide additional data, and I was most interested to hear a lecture by Professor O. E. Radczewski, of the Institut für Gesteinshüttenkunde of Aachen, Germany, describing some recent mineralogical results in this area. The modern electron microscope, with a resolving power approaching 4 Å, opens up new possibilities of study of minute inclusions and even perhaps lattice distortion. Further, with techniques for obtaining diffraction patterns from grains of submicron size, more information on the structure and composition of these minute inclusions is possible.

Thus, in increasing measure, any physicochemical analysis of the trace-element problem will not be frustrated by positional assumptions. In the study of polymorphs, for example, we can now be much surer about the significance of impurity effects on stability. Will this new information lead to modification of existing concepts of the factors governing minor element distribution? Almost certainly it will. We are all troubled from time to time by apparent tolerances and intolerances of the crystal lattice as a solvent.

With a few noteworthy exceptions, the discussions of factors governing element distribution and fractionation in geological processes have been primarily concerned with the thermodynamic properties of the solid phase containing the trace element. That all such treatments should be less than satisfying is obvious when the processes involved are considered in detail. Shaw (1953) has been one of the few to stress that solid-fluid equilibrium must be considered. It is not enough that an atom or ion be strongly bound in a crystal lattice; we must assess how it is bound in the phase from which it is separating. In short, most of the rules used in discussing the substitution of ions in a lattice are relevant only if the formation process is of the type

$$X^{+}_{\text{gas}} + YZ_{\text{solid}} \rightleftharpoons Y^{+}_{\text{gas}} + XZ_{\text{solid}} \,.$$

As soon as we introduce any medium more complex than the gas phase, an aqueous fluid or silicate melt, then any analysis of value must involve an estimate of thermodynamic functions for the steps of a cycle such as

$$\begin{array}{ccccc} X^{+}_{\text{gas}} + & Y^{+}_{\text{gas}} + Z^{-}_{\text{gas}} & \rightarrow & XZ_{\text{solid}} & + Y^{+}_{\text{gas}} \\ & \uparrow & & & \downarrow \\ X^{+}_{\text{fluid}} + & YZ_{\text{solid}} & \rightleftharpoons & XZ_{\text{solid}} & + Y^{+}_{\text{fluid}} \,, \end{array} \tag{1}$$

and the extent of reaction (1) must depend on (*a*) the relative lattice energies of the solids and (*b*) the relative solvation energies of the ions. Our problem is unfortunately not simplified by the knowledge that factors that lead to large lattice energies also lead to large solvation energies. And, since heats of fusion of most solids are small, this implies that liquid-state binding forces are very similar in magnitude to solid binding forces. A simple example of such competition is indicated by the comparison below of the reactions

$$\text{Li}^{+}_{\text{aq.}} + \text{NaF}_{\text{solid}} \rightleftharpoons \text{Na}^{+}_{\text{aq.}} + \text{LiF}_{\text{solid}} \tag{2}$$

and

$$\text{Li}^{+}_{\text{aq.}} + \text{NaI}_{\text{solid}} \rightleftharpoons \text{Na}^{+}_{\text{aq.}} + \text{LiI}_{\text{solid}} \,. \tag{3}$$

For reaction (2) $\Delta H^{\circ} = -1.1$ kcal., $\Delta G^{\circ} = -2.7$ kcal. and $\Delta S^{\circ} = +5.5$ cal. For reaction (3), $\Delta H^{\circ} = +13.25$ kcal., $\Delta G^{\circ} = +11.1$ kcal. and $\Delta S^{\circ} = +7.0$ cal. Under standard conditions reaction (2) will tend to the right and (3) to the left, so that LiF will tend to coexist with Na^{+} ions and NaI with Li^{+} ions in aqueous solution. The different trends of these similar processes are a reflection of relative values of lattice energies and hydration energies.

If we are to achieve a real understanding of distribution and fractionation processes, we must obtain data for the complete process. Improvement of the empirical short cuts of the past cannot be satisfying until this is done for a large number of cases. The requirements for such analysis are the study of trends in heats and entropies of solvation in typical geological reaction media. Much of these data may already lie buried in existing phase diagrams, but spectroscopic and electrochemical investigation of melts is essential. Similarly, we need more precise data on heats of mixing in solids, and the entropy of mixing may frequently be amenable to calculation (for an approach see Bradley, 1962).

Many cases of great mineralogical and petrological significance involve transition metals. As we know, these are elements characterized by an incomplete shell of *d* electrons inside an outer *s* electron shell. For example, the 4*s* electrons of the iron atom are about 4.5 times as far out from the nucleus as the 3*d* electrons (Coulson, 1961). Thus to a first gross approximation, the 3*d* electrons may be considered to occupy an inner position, remote from bonding electrons. In the free atom, the five *d* electron levels are of equal energy or degenerate. But it was recognized by Bethe in 1929 that this degeneracy would be removed in an electrostatic field of a group of negative ions surrounding the transition metal ion. The theory covering these effects, crystal or ligand-field theory, developed at this time by Bethe, has been rejuvenated over the past five years or so and has provided a rational interpretation of a great number of observations on optical and magnetic properties and the heats of formation of transition metal compounds (Orgel, 1960).

Although the chemical literature has contained many reviews dealing with ligand-field theory in the past few years, little application has yet been made in our science. However, we can hardly afford to neglect a theory dealing with over 40 per cent of the constituents of the earth. The ligand-field approach is simple to visualize. The wave functions for *d* electrons give electron distributions for the five *d* orbitals, as shown in figure 2. Consider an *xy* plane in a crystal with the halite structure containing a transition metal in octahedral coordination (fig. 3). If we place negative charges in the form of anions or oriented dipoles along the *x*- and *y*-axes, clearly an electron in the $d_{x^2-y^2}$ orbital suffers more repulsion than an electron in the d_{xy} orbital. Thus, in the field, the degeneracy of the *d* levels is destroyed and the five orbitals are split into two sets, as indicated in figure 4. The difference in energy between these levels, Δ, can be determined by optically exciting electrons from lower to upper level, and, since Δ is commonly in the range 1–4 electron volts, the transitions occur in or near the visible region of the spectrum and are responsible for the color of transition metal compounds. The complexity of the observed spectrum depends on the symmetry of the ligand field and the number of *d* electrons in the metal. The position of the

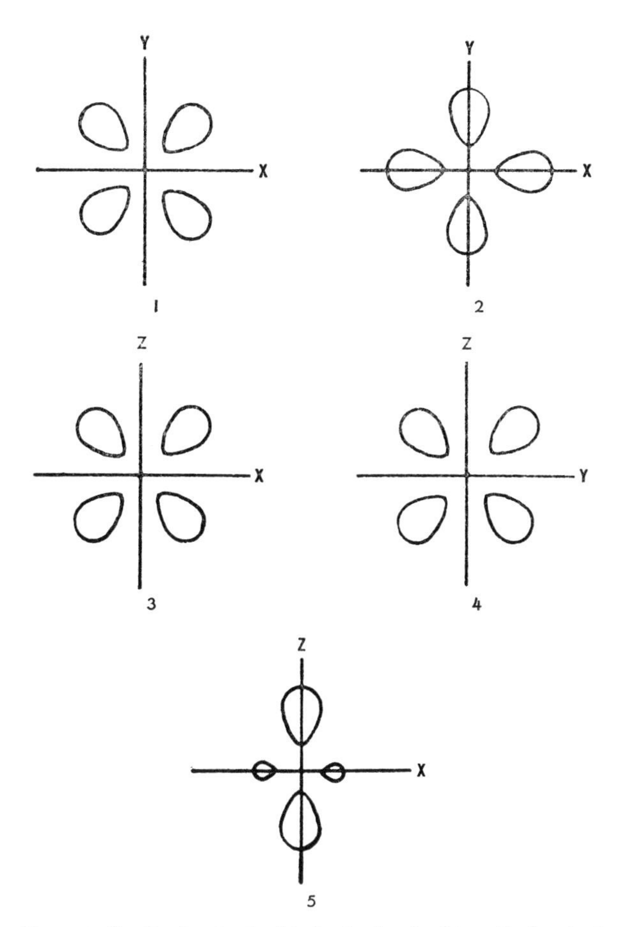

FIG. 2.—Electron distribution in d orbitals: 1, d_{xy}; 2, $d_{x^2-y^2}$; 3, d_{xz}; 4, d_{yz}; 5, d_{z^2}

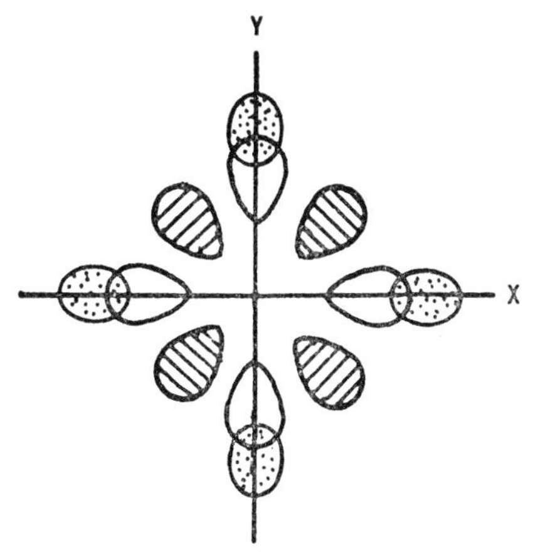

FIG. 3.—Representation of the x–y plane of a transition metal ion in an octahedral field. The d_{xy} orbital is cross-hatched, the ligands or anions are stippled.

absorption bands depends on the strength of the field or the electronegativities and distances to the ligands. Thus, studies of absorption spectra can supply a great deal of information on crystal chemistry and oxidation states. As an example, R. H. Clarke and I have examined the color of the mineral benitoite, $BaTiSi_3O_9$. If this mineral contained all Ti^{4+} with no *d* electrons, it should be colorless. The absorption spectrum indicates that it does not contain Ti^{2+} and that Ti^{3+} is only a remote possibility. Therefore, the color is not due to transitions within Ti ions and is probably caused by defects, such as *F* centers.

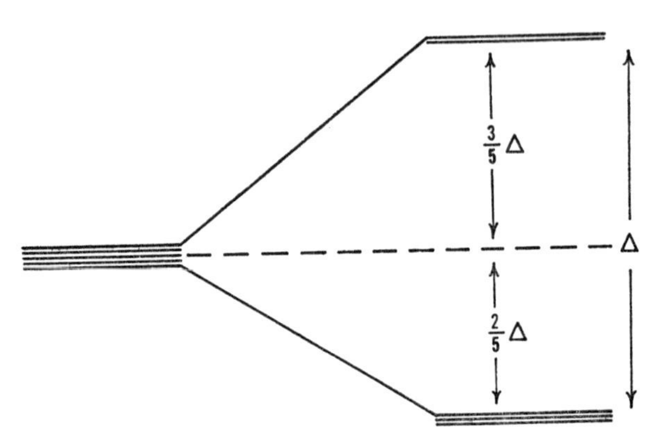

FIG. 4.—Energy diagram for the five *d* orbitals in an octahedral field. In the absence of an external field, the five levels are degenerate. In the octahedral field they are split as shown.

The separation of split *d* levels in a field can lead to electronic stabilization of a compound. If the octahedral field is again considered, it can be shown that the upper *d* levels can be considered to be $\frac{3}{5}\Delta$ above the ground state and the lesser levels $\frac{2}{5}\Delta$ below. If we consider transition metal ions with 1–10 *d* electrons (i.e., Ti^{3+} to Cu^{+}), the relative stabilization associated with the crystal field is

zero for 0, 5, 10 *d* electrons ,

0.4Δ for 1 and 6 *d* electrons ,

0.8Δ for 2 and 7 *d* electrons ,

1.2Δ for 3 and 8 *d* electrons ,

0.6Δ for 4 and 9 *d* electrons .

These values add to thermochemical quantities significant terms of the order of an electron volt. I shall return to other aspects of this below, but in the context of element distribution it should be stressed that, whenever a transition metal ion is transferred from one environment to another, for example, from a silicate melt to a crystal, changes in Δ associated with changes in the environment may lead to significant electronic energy changes not expected from consideration of any ideal ionic or covalent model. For example, the stabilization of an ion in a melt should be less than in a solid; thus, other factors being equal, the number of d electrons in the transition metal ion and the value of Δ in the solid may play a large part in determining competition for a lattice site of given coordination. Williams (1959) has briefly considered this aspect of fractionation in the Skaergaard and has concluded that this factor may play a dominant role. To place such possibilities on a firm foundation, we require much more information on Δ values for transition metal ions in liquids, solids, and solid solutions.

In summary, we may say that we can carry out today, in a way never before possible, the detailed study of both abundance and distribution of elements in rocks and minerals. Further, we are aware of most of the factors involved in fractionation and distribution. But, at the present, exact prediction of these factors is impossible. The empirical rules often used rest on shaky foundations and work only when they do because they have been forced to fit a certain set of results. Given more values of relevant quantities, trends in these should eventually allow reasonable predictions. These quantities must include the thermodynamic functions involved in all types of solvation and solution reactions, and, where transition metals are involved, spectroscopic data are essential for adequate interpretation.

Phase Transitions and Reactions

A great deal of experimental work aimed at elucidation of the physicochemical history of the crust of the earth has been based on the assumptions (*a*) that, in general, chemical conditions reflecting the highest pressures and temperatures are preserved and (*b*) that local equilibrium was attained at that time. A host of observations separated in place and time supports the assumptions and indicates their limitations. These assumptions have also led to the belief that, with intelligent experiment, we can outline physicochemical conditions for petrological reactions. At the outset, any experimenter must be aware of one variable that he can seldom duplicate. This variable is time, and an outsider examining our problems might thus expect that he would find an enormous proportion of our effort involved in the study of reaction rates. He would be disappointed to inspect our literature and would not be surprised to find that the significance of many experiments was most limited.

In this respect, the most amenable problems would involve such phenomena as volcanism, where the time problem is minimal, and here one must congratulate the workers of the Carnegie Institution, where so much early work involved solid-melt equilibria. But in studies in the subsolidus region, a region covering most of the crust and mantle, rate problems are real, and I will return to these below.

If we assume that we can overcome kinetic problems, and in many and perhaps the majority of cases we can, we still must face the problem of where we are going in our phase studies. One of the great contributions that experiment has made in the past is that it has suggested good models of complex phenomena based on simple systems. The specification of exact numbers for complex systems is limited on account of the variety of chemical and physical variables. Is a geologist much interested, and does our science advance much, if I can say that a certain green schist formed at 325° ± 1° C. and 4,215 bars as compared with the statement that it formed at 300° ± 25° C. and 4,000 ± 1,000 bars? Can such precision of data be usefully applied? Certainly, if a given mineral needs a pressure of formation in excess of 50,000 bars, this is geologically important. But how well must we know this figure? On the other hand, the simple model systems of two and three components studied by the Carnegie Institution in its early work caused a revolution in thinking about igneous petrology.

Some time ago, while speaking with Dr. Stuart Agrell, of Cambridge, I suggested that a certain phase boundary was perhaps 50° C. too high. His reply was something like "So what?" His point is well made. In the initial stages of experimentation there is every reason to achieve the maximum degree of precision and to try all possible approaches to see what works best. But before we grind out system after system we must answer the question of what we wish to achieve.

One can take several views of this problem. First, we are amassing data on systems that, if reliable, can provide thermodynamic data on inorganic systems of high precision. Some time ago, I indicated the desirability of obtaining more precise direct calorimetric data on mineral systems. But, if good direct studies are possible—as in many cases they are—the precision in the fix on free energies is most precise. For example, in the reaction

$$\text{quartz} \rightarrow \text{tridymite}\ ,$$

Tuttle and Bowen have fixed the equilibrium transition temperatures with a precision of ±10° C. As the entropy of this transition found by them is about 0.5 cal/mole deg, this represents fixation of relative free energies to within ±5 cal/mole. This uncertainty is far superior to any direct calorimetric estimate. A combination of calorimetric estimation of entropies and direct study of equilibrium to provide free energies and heats may produce very satisfactory thermodynamic functions for inorganic materials over a wide *P-T* range.

If this becomes part of our objective, the approach to phase studies changes. For the accurate determination of thermodynamic functions, high precision of a few points is more valuable than the exploration of a broad system. Only reactions convincingly shown to be reversible are significant. If synthesis alone is the objective, then the approach is far less rigorous.

In the petrological context, modern rock experimenters will perhaps become more concerned with limited reactions when field observations have shown them to be critical—or with whole rock reactions. There may be difficulties with these, but, if approached as rate studies in both directions, valuable limits on conditions and processes may result.

There is, however, a vast, relatively untouched area in this part of our science. For instance, we should attempt to explain within a reasonable framework such things as the range of stability of polymorphs and the compatibilities and incompatibilities of assemblages (e.g., why is albite more stable than nepheline and quartz; fayalite-quartz more stable than ferrosilite, etc.?). As we have discussed previously (Fyfe, Turner, and Verhoogen, 1958), we can frequently guess entropy changes with reasonable precision. Our problem lies in estimating heats of reaction, and when we discuss these we are concerned with the strength of chemical bonds. How seldom do we see these basic questions discussed? It is easy, and in part true, to say that such predictions are beyond any theoretical attack. But let me remind you of the enormous progress made in determining factors that control rates of organic reactions. This began with a multitude of experiments; these led to empirical rules, and, today, molecular-orbital theory has placed all on a sound basis and has produced valuable and reliable predictions. Our problem becomes less severe if we concern ourselves not with absolute values for thermodynamic parameters but only with differences, which are the thermodynamically significant quantities. In the geophysical and astrophysical realms, experiment will always be a long way from reproducing all conditions. To extrapolate safely, we need the best possible theoretical analysis.

May I take two examples of empirical approaches? There has been immense interest in the synthesis and finding in natural materials of stishovite, SiO_2 with the rutile structure. To most of us, including geophysicists, this discovery was a surprise. But its probable region of existence could have been predicted with little difficulty.

Let us ignore coesite and consider the transition of quartz to stishovite. We may base our comparison on the known heats of formation of rutile and quartz. The molar volume of quartz is 22.7 cm^3/mole. We know that silicon is smaller than titanium, but let us assume that SiO_2 goes into the rutile structure with the same molar volume as rutile. This SiO_2 would have a molar volume of 18.8 cm^3/mole. These figures tell us that at some pressure the transition from the quartz structure to the rutile structure is very likely to occur. Next, we must obtain some better estimate of bond distances, and we might examine silicon analogs of titanium-bearing ions or compounds with identical coordination numbers. The structures of some of the hexafluorides of both are known, and the Ti-F and Si-F distances in TiF_6^{2-} and SiF_6^{2-} are 1.82 ± 0.02 Å and 1.71 ± 0.02 Å, respectively. Thus, since the mean Ti-O distance is rutile in about 1.96 A, we would anticipate a Si-O distance in stishovite of the order of 1.84 Å. With this figure we would anticipate a density of stishovite near 4.0.

To obtain an estimate of the heat of transition, we might first consider lattice energies. We know the Madelung constant for the rutile lattice, and, with our estimated Si-O distance and by comparison with the known thermochemical lattice energy of rutile, we would estimate that the heat of transition of quartz to stishovite might be in the range 78 ± 53 kcal/mole. The large uncertainty is a reflection of the large lattice energies (*ca.* 2,500 kcal.) and the uncertainty in distances. We can guess ΔS (see Fyfe, Turner, and Verhoogen, 1958), and with this ΔG° will be of the order of 80 ± 53 kcal/mole. If we use the known ΔV of 9 cm^3/mole (Sclar *et al.*, 1962), the transition pressure might be in the range 120,000–500,000 atm. The uncertainties are colossal, but stishovite is a feasible compound.

We can approach the problem in other ways. The ΔH's of the reactions

$$Ti_{solid} + 6F^-_{aq.} \rightarrow TiF^{2-}_{6\ aq.}$$

and

$$Si_{solid} + 6F^-_{aq.} \rightarrow SiF^{2-}_{6\ aq.}$$

are known and are −555.1 and −558.5 kcal/mole, respectively. If we analyze these reactions, we see that most factors involved are comparable with the reactions

$$Ti_{solid} + O_{2_{gas}} \rightarrow TiO_{2_{solid}}\ (\Delta H = -\ 218\ \text{kcal/mole})$$

and

$$Si_{solid} + O_{2_{gas}} \rightarrow SiO_{2_{solid\ (stishovite)}}\ \Delta H = ?$$

The main factors that differ involve the heats of hydrations of the complex fluoride ions. These differences may be estimated very approximately by noting that SiF_6^{2-} is a smaller ion than TiF_6^{2-} and by assuming, therefore, that the correction is of the order of 15 kcal/mole. Thus our guessed value for the heat of formation of stishovite is around −206 kcal/mole, compared with −205 for quartz. All things considered, this is probably a better guess than that from lattice energies, and the figures indicate that the free energy of the reaction quartz → stishovite is quite small. Although there is still uncertainty in the exact position of this phase boundary, the heat of the reaction is probably of the order of +10 kcal/mole. The estimates are reasonable, and, if the heats of formation of the fluorides were more reliable, they might be improved.

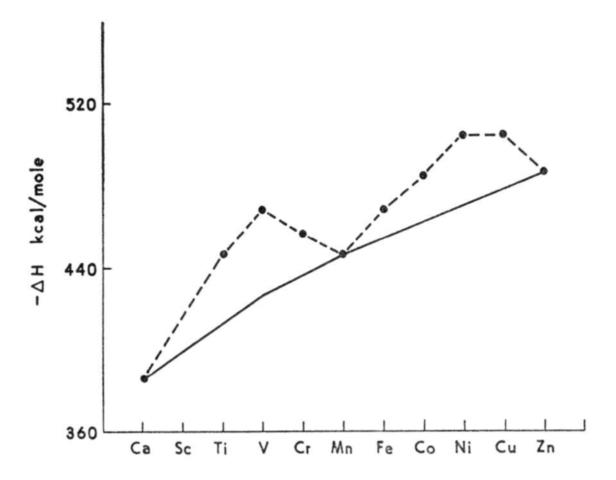

FIG. 5.—Heats of hydration of divalent metal ions. The dashed line represents the measured values; the full line, values after ligand-field correction (after Orgel, 1960).

Another interesting case is presented by Bates, White, and Roy (1962), who have reported data on the transition

$$ZnO_{\text{sphalerite structure}} \rightarrow ZnO_{\text{halite structure}} .$$

Normal zinc oxide is an interesting case, in that, according to the rules of the ionic model and radius ratio considerations, ZnO should have the halite structure with a cell edge of 4.28 Å, exactly as found by Bates *et al.* In this case we would predict the volume change of 2.5 cm^3/mole exactly. Again the problem involves heats of reaction, and one might start by comparing heats of formation of various RO oxides, where R represents cations of similar radius and electronegativities to zinc. It has been known for some time that the heats of formation of transition metal oxides do not increase smoothly with decreasing size of cation as lattice-energy theory would predict. Again ligand-field theory has explained this anomaly. If we compare heats of formation of zinc and transition-metal compounds of comparable structural states (see fig. 5), we find that, after ligand-field corrections, zinc compounds have rather constant relations to the latter. Manganous compounds are of particular interest because, with this ion, there is no field correction. From data in Orgel (1960) we find that the lattice energies of zinc compounds and the hydration energy of the zinc ion always exceed the corresponding manganous compounds by 45–50 kcal/mole. Thus, we may estimate the lattice energy of zinc oxide in the simple cubic structure as

$$U_{\text{ZnO, sphalerite structure}} = 970 \text{ kcal/mole} ;$$

$$U_{\text{MnO, halite structure}} = 920 \text{ kcal/mole} ;$$

$$U_{\text{ZnO, halite structure}} \simeq 965 \text{ kcal/mole} ;$$

$$\Delta H^{\circ}_{\text{transition}} \simeq +5 \text{ kcal/mole} .$$

(Lattice energies from Moelwyn-Hughes, 1957.)

The experimental data of Bates *et al.* also indicate that the ΔH° of this transition must be near 5 kcal/mole. The agreement is excellent, but the exact coincidence of the figures is fortuitous.

Once we accumulate data on many of these transitions, estimates of the differences will become more intelligible. For instance, I think that we can now say (*pace* Fyfe, Turner, and Verhoogen, 1958) with reasonable certainty that MgO will not form a CsCl modification over the range of pressure in the mantle but that CaO certainly will. But other problems remain. Why does ZnO prefer a tetrahedral (sphalerite) structure and BaO an octahedral (halite) structure? Why does the ionic model fail in such simple cases? The answers to these questions must sharpen our views on the whole problem of chemical bonding. The study of such transition energies through phase relations may make a valuable contribution to theories of the solid state.

When *d* electron compounds are considered, large changes in crystal-field stabilization may be important factors in transition energies. Further, spin coupling in a strong field can lead to substantial ion shrinkage, as we see in a comparison of bond distances in MnS_2 and FeS_2, 2.59 and 2.27 Å, respectively. In these cases the ions have the structure

Mn^{2+} (↑) (↑) (↑) 3*d* (↑) (↑)

Fe^{++} (↑↓) (↑↓) (↑↓) ( ) ( )

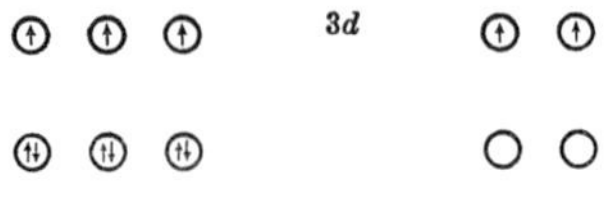

Such electronic transitions can be studied by optical and magnetic methods.

Perhaps I could end this section with a quotation from R. O. Davies

(1962): "Like low-temperature physics, high-pressure physics is a dying subject—not indeed because it is no longer practiced, but because it is being so widely practiced that the element of 'craft' has passed over to technology." There is some truth in Davies' statement, but as long as we keep in mind that all this work can provide new and unique data on the properties of matter, we can keep it alive for a long time to come.

Mechanisms and Rates

As I mentioned above, the experimentalist, while being aware that he frequently cannot duplicate the time factor in geological processes, has chosen to ignore it almost totally. For some phenomena, laboratory times must be of the order of 10^6–10^{10} too short. This difficulty is critical in the planning and analysis of both rate and equilibrium studies, which, in fact, can never be separated.

Why has the rate problem been ignored—is it unimportant, and are most processes of interest rapid? The answers are obviously negative—but the fact that many laboratory reactions produce interesting results in a short time has dulled our awareness of the problem. Recent equilibrium studies have shown the need for much more study. For instance, why is it that, with all the calcium zeolite synthesis during the past few years, one of the most geologically common species, laumontite, has not been produced in an unprejudiced reaction? If aragonite is unstable at low pressures, why is it more easily synthesized at 100° C. than 0° at 1 bar? Why is magnesite almost impossible to produce at room temperatures? Why do igneous rocks persist in low-grade environments, why do minerals retain argon, and why are there so many minerals in the museum? The answers to these and a host of other questions involves the study of rates and mechanisms.

In the study of rate processes, two steps are normally involved. First, a mathematical expression is found that will describe the quantitative features of the process under consideration. The second stage involves the interpretation of this mathematical expression in terms of some reasonable molecular model. The significance of one depends on the other, and I suggest that, if we are to consider extrapolations of the order of 10^8 in time, this dual analysis is fairly essential. With few exceptions, normally involving solid diffusion, I think we do not really understand the mechanism of geological processes. The mechanism of most processes in hydrothermal systems is certainly not known.

If we understand mechanism, then we may more intelligently approach catalysis, and this may relieve some of the critical difficulties. A simple example will illustrate this problem. When amorphous silica is crystallized in an aqueous environment at temperatures at which quartz is the stable modification, the process normally proceeds through the steps

$$\text{amorphous silica} \rightarrow \text{cristobalite} \rightarrow \text{silica-K} \rightarrow \text{quartz}$$

or

$$\text{amorphous silica} \rightarrow \text{cristobalite} \rightarrow \text{tridymite} \rightarrow \text{quartz} .$$

The rates of the over-all processes are not related to rates of solution of the species, and thus growth and/or nucleation must be critical in the rate-determining steps. The rate is most sensitive to pressure or solvent density but is not a direct function of the amount of silica in solution. In the range 300°–500° C., it is rather insensitive to temperature. Near 200° C., the rate is proportional to hydroxyl ion concentration and near 300° C. is proportional to the square of this quantity. All these facts indicate that nucleation is the critical process and is controlled by the concentration of silicate ions. Knowing this, we can catalyze this reaction by factors of at least 10^4.

In almost all heterogeneous processes the rate process is controlled by some of the steps: (*a*) solution rate of reactant, (*b*) nucleation rate of product, (*c*) growth rate of product (controlled by diffusion processes or adsorption processes), and (*d*) recrystallization rate of fine or metastable products.

Rates of solution can normally be studied if we have knowledge of surface areas. Theories of growth have now advanced with the recognition of the importance of dislocations, but geological processes that involve crystal growth at very low degrees of supersaturation still present problems. The recrystallization, or aging process, involves minimal supersaturation; the mechanism of recrystallization might be closely tied to slight and continued deformation in natural systems. It has already been noted by experimentalists that shearing will catalyze sluggish processes, perhaps through the formation of a large number of dislocations.

Nucleation is, however, a problem that we have hardly begun to understand, and it demands detailed study in the most simple cases. If I may borrow an expression from Professor J. S. Anderson, all these processes involve *terra incognita*, the surface of the reacting solid. If we are to understand almost any of our common processes, we require knowledge of solution species and of surfaces. The field is wide open, and the problem is so critical to us that we should try to make a large basic contribution.

Mechanism in solid systems involves the study of every aspect of the defect lattice and the properties of electrons in solids. In the past we have missed (in this area) many obvious clues beautifully displayed in the colors of minerals. Even today, we hardly understand these colors, but, as future work will show, there may be some quite new electron exchange phenomena related to anomalous absorption seen in minerals. The study of mobility of elemental and subelemental species should be an important approach to the problem of reactivity.

I might briefly indicate one reaction that seems to be of considerable interest to geologists at the present time, one that Brown, Fyfe, and Turner (1962) have studied in some limited aspects. This is the reaction:

$$\text{aragonite} \rightarrow \text{calcite} .$$

The problem that concerned us was to explain how metamorphic aragonite, crystallized in its stability field, could survive a prolonged unloading process in the calcite stability field. From studies of transformation times of a powdered sample, we were able to fit a rate law, and from this rate equation "apparent" activation energies can be deduced, and extrapolation to longer times is possible. For many geological purposes one must consider extrapolations in time of the order of 10^6. Chaudron (1952) had carried out similar rate studies and found, for various samples, activation energies varying from 45 to 80 kcal. If this range is real, extrapolation is dangerous, to say the least. Recently, Miss Margaret Lang has been carrying out further studies on the rapid and reversible analog of this transition in KNO_3. In this case (and to some extent in the case of aragonite) the transition can be followed visually in single crystals, and a number of features have emerged. The primary nucleation step is both structure sensitive (sensitive to the history of the crystal) and habit sensitive. Once nucleation has occurred, the rate involves a linear boundary advance in the direction of the *c*-axis, a process with zero-order kinetics. The apparent rate law, the rate constants, and activation energies are thus both structure and habit sensitive. Before long-range extrapolation is considered, these factors must be assessed for the actual material in question. Thus I wish to stress the importance of some detailed knowledge of how a given process occurs.

Solutions—Mixtures

In this lecture I can touch on only a few features of this vast geochemical field. Practically every phase that we deal with, and every mechanism that we propose, will at some stage require consideration of solution or mixing—a phenomenon dominated by the simple thermodynamic consideration that mixing involves an increase in entropy.

The geochemistry of the oceans and sediments is, in many of its most fundamental aspects, a problem of ionic equilibria in a complex electrolyte-nonelectrolyte mixture. Our understanding of this field has advanced a long way, thanks to a vast effort by electrochemists. In this problem there are processes requiring knowledge of ion activities, complexing, ion-pairing, and oxidation-reduction. Garrels and his school have used and amplified much of this type of data, but a problem arises in the geochemical context that is critical. To make predictions, we are frequently confronted with the need of knowledge of single ion activities in a mixture. These activities are not simple to measure directly—some would say that they are impossible to measure. Here again I think we can make a fundamental contribution by the study of solubilities and precipitation reactions in synthetic media gradually approaching the complexity of natural media. Much is being done, but much more should be done before we understand some common processes. The by-product will be an increased knowledge of the thermodynamics of mixtures and nucleation and growth processes.

When we move from the low-pressure–low-temperature aqueous environment to metamorphic and igneous aqueous solutions, the state of ignorance increases *n*-fold. We have known for a very long time that almost all metamorphic processes require an aqueous solvent. The ore-forming processes require a solvent—inclusions in the minerals tell us this (Roedder, 1962)—yet what do we know of these solutions? In general terms, we have three great markers in this field: the work of Kennedy on *P-V-T* relations in water, the work of Kennedy and others on the system SiO_2-H_2O recently extended to the upper critical end point (1962), and the work of the German electrochemist E. U. Franck (1961) on the behavior of some classical strong electrolytes in high-temperature aqueous solutions. Against such concrete contributions, we can list a host of vague discussions and extrapolations that have served only to keep a sickly patient alive. We do not, I think, have any adequate picture of the solution chemistry of a single metal silicate, and we do not even fully understand the species present in SiO_2-H_2O solutions. This is a very old problem, but every significant piece of data added is fundamental and leads to the exciting day when we can discuss intelligently 1,000° C. aqueous chemistry.

Let us look at the broad aspects of this problem. Franck has shown us that over much of the range of possible temperatures and pressures that might occur within the crust, electrolytes are all weak with dissociation constants in the range 10^{-2}–10^{-6}. Further, salts such as KCl produce dominantly molecular solutions which will not be neutral but, depending on physical conditions, may be either acid or alkaline. The self-dissociation of water has been measured over a most limited range of conditions, but estimates in other regions may be reasonable. Brewer (1951) has stressed that the vapor phase in equilibrium with solids increases in molecular complexity as temperature increases. We thus are confronted with two rather unfamiliar problems, the formation of molecular salt solutions and the formation of complex polymeric vapor species. The possibilities to which these considerations lead are, in my opinion at least, beyond the range of any adequate predictions. Measurements must be made, and techniques are available, that will provide some of these data. A great deal can be obtained from the study of a combination of solubilities, solubilities in mixtures, conductance, and even absorption spectra, which can be used at least in some cases to 500°. This great gap in our knowledge must be filled—or are we going to wait for the chemists again?

If we are to understand silicate phase reactions in aqueous media, we must obtain data on silicate solutions. As I have mentioned previously, quartz crystallization rates appear to depend on silicate ion species and can hence be catalyzed. Do all types of silicates produce the same solution species rapidly? Why are some silicate crystals notoriously difficult to nucleate; is this a reflection of a low concentration of some critical solution species? I think the answers to these questions are of greater importance than knowing that the boundary of the greenschist facies is precisely at 225° C.

Discussions of the effect of solid solutions on mineralogical processes have frequently assumed, in the absence of contradictory data, that such solutions are ideal. Many of the processes with which we are concerned have embarrassingly small free-energy changes, and hence our assumption becomes dangerous. Almost by definition, transition-metal solid solutions will not be ideal. Crystal-field stabilization must change with composition, temperature, and pressure. Mr. Roger Burns, working in our laboratory, has been able to measure the magnitude of this change in the olivine series at room temperature and has found a significant increase in stabilization of the ferrous ion as the magnesium ion concentration in the olivine increases. This mixing term changes by about 1,000 calories with composition, a value that will decrease at high temperature and increase with pressure. We do not yet understand the full significance and magnitude of such effects, but the chemistry of many solid solutions and problems of stability (e.g., $FeSiO_3$) may require this information for explanation. This has already been shown in the spinel series (see Orgel, 1960). The same studies increase our detailed knowledge of binding forces in solids.

Conclusions

I have touched on a number of obvious gaps in geochemical knowledge; many more could be mentioned. Surface reactions must be of great significance in sedimentary processes; our knowledge of the solid-state physics of minerals is vague. Irreversible thermodynamics clearly has applications in our science, for the concept of local equilibrium implies steady-state boundary conditions. A start has been made, for example, by the work of Jaeger (1961) and Shimazu (1958, 1960).

Throughout this discussion I have stressed the need for more basic data. In many areas, experiment is needed to guide in the extension of physicochemical principles. In general, fundamental geochemical advances have waited on the physicist and chemist to break the ground—in too few cases have we led. We are, thus, borrowers and, if we are not careful, will find ourselves in a state of scientific bankruptcy.

If we are to proceed rapidly, it is becoming more and more important that the student of geology be exposed to the maximum amount of physics and chemistry from physicists and chemists. Unless this is done, we cannot hope to speed up infiltration rates of knowledge, and the language barrier of technological jargon will become an increasing difficulty. Only students with this type of training will have the tools to select the basic problems and attack them realistically in the future.

References Cited

Bates, C. H., White, W. B., and Roy, R., 1962, New high-pressure polymorph of zinc oxide: Science, v. 137, p. 993.

Bradley, R. S., 1962, Thermodynamic calculations on phase equilibria involving fused salts, pt. 2, Solid solutions and applications to the olivines: Am Jour. Sci., v. 260, p. 550–555.

Brewer, L., 1956, Undiscovered compounds: Jour. Chem. Education, v. 35, p. 153–156.

Brown, W. H., Fyfe, W. S., and Turner, F. J., 1962, Aragonite in California glaucophane schists, and the kinetics of the aragonite-calcite transformation: Jour. Petrology, v. 3, p. 566–582.

Chaudron, G., 1952, Contribution a l'étude des réactions dans l'état solide cinétique de la transformation aragonite-calcite: International Symposium on the Reactivity of Solids, Gothenburg, p. 9–20.

Davies, R. O., 1962, Aspects of solid-state physics: Nature, v. 194, p. 619.

Denbigh, K. G., 1955, The principles of chemical equilibrium: Cambridge, Cambridge University Press.

Franck, E. U., 1961, Überkritisches Wasser als elektrolytisches Lösungsmittel: Angewandte Chemie, v. 73, p. 309–322.

Fyfe, W. S., Turner, F. J., and Verhoogen, J., 1958, Metamorphic facies and metamorphic reactions: Geol. Soc. America Mem. 73, 259 p.

Goldschmidt, V. M., 1937, The principles of distribution of chemical elements in minerals and rocks: Jour. Chem. Soc. London, p. 655–673.

Jaeger, J. C., 1961, The cooling of irregularly shaped igneous bodies: Am. Jour. Sci., v. 259, p. 721–734.

Kennedy, G. C., Wasserburg, G. J., Heard, H. C., and Newton, R. C., 1962. The upper three-phase region in the system, SiO_2-H_2O: Am. Jour. Sci., v. 260, p. 501–521.

MOELWYN-HUGHES, E. A., 1957, Physical Chemistry: London, Pergamon Press.

ORGEL, L. E., 1960, An introduction to transition-metal chemistry: London, Methuen & Co.

ROEDDER, E., 1962, Ancient fluids in crystals: Scient. American, v. 207, p. 38–47.

SCLAR, C. B., YOUNG, A. P., CARRISON, L. C., and SCHWARTZ, C. M., 1962, Synthesis and optical crystallography of stishovite, a very high pressure polymorph of SiO_2: Jour. Geophys. Research, v. 67, p. 4049–4054.

SHAW, D. M., 1953, The camouflage principle and trace-element distribution in magmatic minerals: Jour. Geology, v. 61, p. 142–151.

SHIMAZU, Y., 1958, A theory of physico-chemical reactions under a non-uniform field and its application to a study of thermodynamical state of the earth: Jour. Earth Sci., Nagoya University, v. 6, p. 31–51.

——— 1960, A role of water in metamorphism as illustrated by some reactions in the system MgO-SiO_2-H_2O: *ibid.*, v. 8, p. 86–92.

TUTTLE, O. F., and BOWEN, N. L., 1958, Origin of granite in the light of experimental studies in the system $NaAlSi_3O_8$-$KAlSi_3O_8$-SiO_2-H_2O: Geol. Soc. America Mem. 74, 151 p.

WILLIAMS, R. J. P., 1960, Deposition of trace elements in basic magma: Nature, v. 184, p. 44.

FRED A. DONATH

Fundamental Problems in Dynamic Structural Geology

STRUCTURAL GEOLOGY is the field of study concerned with the causes, the mechanisms, and the effects of geologic deformation. The great variety of topics in structural geology has fostered an even greater variety of problems. Some of these problems can be resolved through improved methods of observation or more careful observation, others will always remain in the realm of hypothesis, and some are purely provincial. A few are truly fundamental.

The structural geologist seeks to answer two basic questions: "How is a structure related in space and time to the stress responsible for it?" and "Why has this particular structure developed?" The determination of the relation between structure and stress distribution is the heart of structural analysis; it is an explanation of the geometry and movements of deformation in terms of the stresses and forces responsible for observed relations. The basis for understanding why a particular structure has developed lies in an appreciation of the effects of environmental and rock factors on the mode of deformation.

Four approaches can be followed to provide understanding of the relation between a structure and its deformational environment—the field approach, the experimental approach, the theoretical approach, and model studies. Of these, one commonly chooses between two alternatives: (1) to work with real geologic materials and collect facts from the field or from experiment or (2) to work with theory or models and, hence, with hypothetical or ideal materials. The two alternatives cannot be considered mutually exclusive, for results obtained from theoretical or model studies must eventually be related to the behavior of geologic

Dr. Donath was born in Minnesota and attended the University of Minnesota and Stanford University, receiving the Ph.D. degree in 1958 from the latter institution. He joined the faculty of Columbia University in 1958, and he is currently an associate professor of (structural) geology. His primary interest has been the analysis of geologic deformation based on the principles of mechanics and on laboratory and field investigations.

materials. Moreover, conclusions drawn from experimental, theoretical, or model studies must be consistent with field evidence. But the fact that they are consistent does not necessarily make them correct. Similarly, conclusions based on field relations must be consistent with experimental evidence or basic physical laws. These points cannot be overemphasized, for their neglect can lead to serious misconceptions.

From experimental studies a great deal is being learned about the deformational behavior of rocks under different physical conditions. From field evidence the mechanisms and geometry of deformation can be related to differences in rock behavior. What are not known are the specific physical conditions responsible for a given geologic structure and the strain or stress distributions actually existing in the structure. These gaps in present knowledge represent the essential link between observation and theory.

The writer considers the most fundamental problems in dynamic structural geology to be those concerned with establishing the relation between structure and stress distribution and with determining the effects of environmental and rock factors on the mode of deformation. Accordingly, this paper will treat three aspects of the subject: first, the role of ductility and anisotropy in determining the mode and geometry of deformation; second, specific techniques for determining strain or stress distribution and ductility in naturally deformed rocks; and, third, the use of theoretical and model studies.

Importance of Ductility and Anisotropy

Environmental Factors and Ductility

Rocks deform because they are not in equilibrium with their environment. The manner in which they deform is primarily a function of ductility—the ability of a rock to flow without fracture. A rock having low ductility undergoes negligible flow before it fractures and is said to be brittle. A rock having high ductility can flow continuously without fracture. Ductility is not constant for a given rock; it is a function of environmental conditions, such as confining pressure, temperature, and strain rate (cf. Handin and Hager, 1957, 1958; Paterson, 1958; Heard, 1960, 1963; Griggs and Handin, 1960).

The importance of ductility in determining the mode of deformation has been nicely demonstrated by Paterson (1958). Plate I illustrates the transition from brittle to ductile deformation as a function of confining pressure for a coarse-grained, pure marble. Cylindrical specimens of the marble were placed in a pressure vessel and deformed under different confining pressures. Longitudinal extension fractures and simple shear fractures developed in specimens of the marble deformed at confining pressures below 100 bars, conjugate shear fractures and conjugate shear zones were favored at pressures from 200 to 300 bars, and distributed shear and uniform flow prevailed for confining pressures above 450 bars. A pressure of 400 bars is the depth equivalent of approximately one mile. Paterson concluded that for this particular rock one mile of overburden would be sufficient to induce distributed shear or flow rather than a simple shear or shear zone.

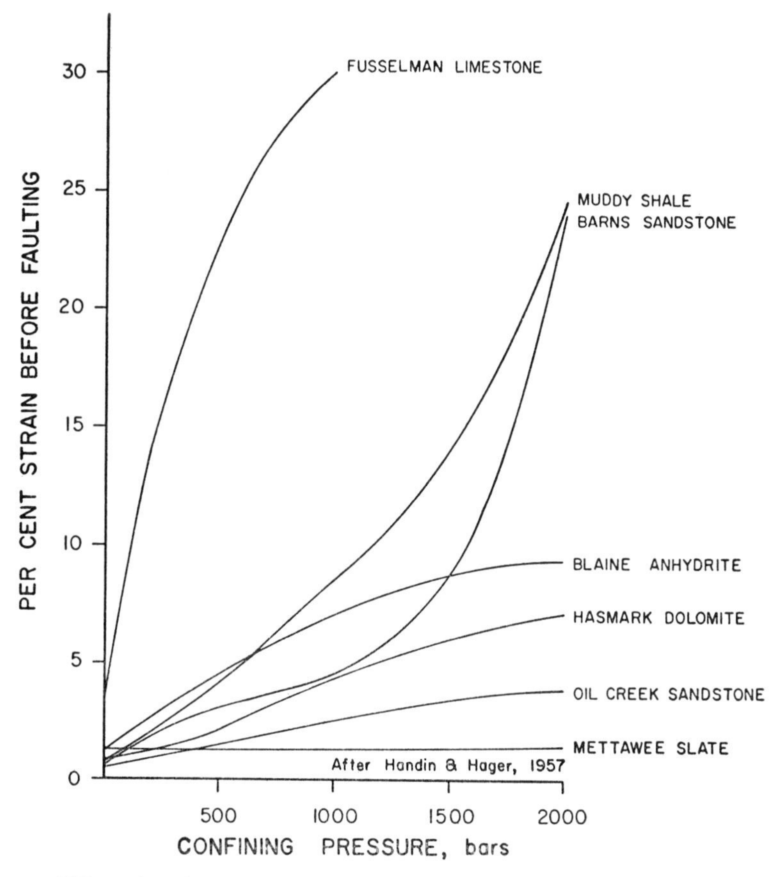

Fig. 1.—Effect of confining pressure on ductility. (Data from Handin and Hager, 1957.)

Experimental work attempts to assess the relative importance of certain variables that affect the ductility, and hence the mode of deformation, of common rock types. For purposes of comparison, ductility is expressed as per cent strain before faulting. The effect of confining pressure on the ductility of several rock types is shown in figure 1; data are from Handin and Hager (1957). The Mettawee slate and Oil Creek sandstone showed negligible flow before fracture and were brittle at all confining pressures to 2,000 bars, a depth equivalent in the earth of about five miles. Minor-to-significant flow preceded fracture in Hasmark dolomite and Blaine anhydrite subjected to 2,000 bars confining pressure, and their behavior at that pressure can be described as moderately brittle to moderately ductile. The Barns sandstone proved brittle at

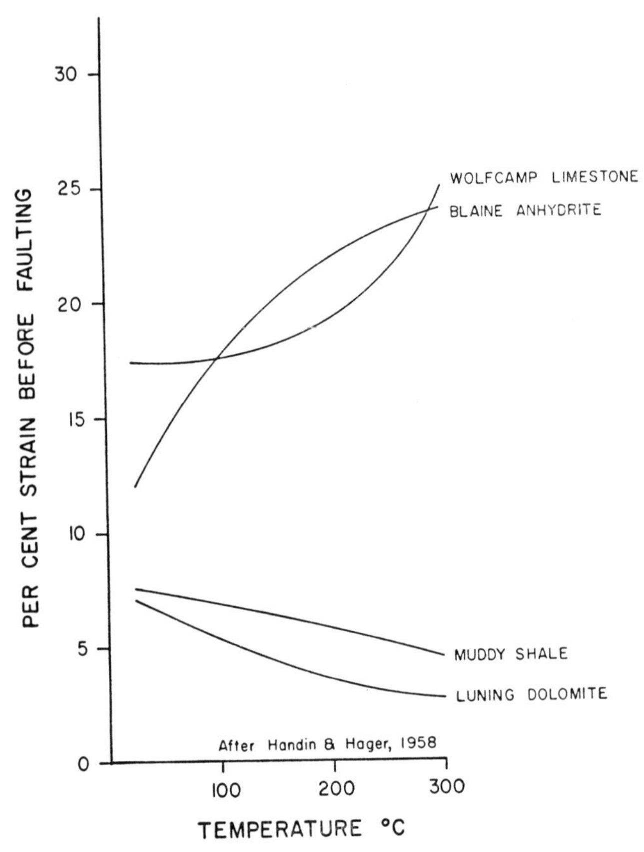

Fig. 2.—Effect of temperature on ductility at 2,000 bars confining pressure. (Data from Handin and Hager, 1958.)

confining pressures below 1,000 bars but became very ductile at 2,000 bars. The Muddy shale became increasingly ductile with higher confining pressure, and the Fusselman limestone was very ductile even at 1,000 bars confining pressure—flowing continuously to 30 per cent strain. One may conclude that the ductility of most rocks is increased with higher confining pressure, though the effect differs from one rock to the next.

In general, higher temperature also increases ductility. Apparently, however, higher temperature can lower slightly the ductility of certain rocks, or lower it slightly before increasing it (fig. 2). In tests at 2,000 bars confining pressure the Muddy shale and Luning dolomite showed slight decreases in ductility with temperature increase to 300° C. Handin (personal communication, 1962) has pointed out that the results for shale would be misleading if high temperatures permit more compaction to occur under the hydrostatic pressure imposed on the specimen while the experiment is being set up. The specimen would presumably then compact less under subsequent application of differential stress and thus would show a lower per cent strain before faulting. The behavior of the dolomite is enigmatic. By lowering the critical resolved shear stress for intracrystalline gliding, high temperature normally induces greater ductility in rocks in which this is an important mechanism of flow. The Wolfcamp limestone and Blaine anhydrite both showed large increases in ductility with increased temperature, as might be expected.

According to Handin, pressure effects always override temperature effects for depths less than about 30,000 feet. Little experimental work has been done to evaluate the effects of strain rate, but recent work by Serdengecti and Boozer (1961) and Heard (1963) indicates that it, too, strongly influences ductility. Lowered strain rates normally allow a greater amount of strain before faulting or loss of cohesion.

Thus, experimental work shows that the ductility of a rock can change appreciably, depending on the environment, and, moreover, that ductility determines to a large extent the mode of deformation and, hence, whether fracturing, faulting, folding, or other type of deformation will develop under a given set of conditions. Because different rock types are affected differently, certain rocks in the same environment may fracture at the same time that others flow. An appreciation of the role of ductility in geologic deformation provides the basis for understanding why a particular structure has developed and emphasizes the need for determining ductility in the field.

EFFECT OF ANISOTROPY

Whereas ductility controls the *mode* of deformation, anisotropy commonly controls its *geometry* (Donath, 1961, 1962*b*). Layering, cleavage, schistosity, and other foliation constitute types of planar anisotropy that can alter appreciably the geometric relations between structure and stress distribution. In a homogeneous, isotropic material, for example, faults would normally develop at approximately 30° to the direction of maximum compression. Figure 3 shows the effect of cleavage on the angle of faulting in a slate experimentally deformed at several different confining pressures. Both the inclination of anisotropy (cleavage) and the inclination of fault are given with respect to the direction of maximum compression. Data points along the sloping line indicate that faults developed parallel to cleavage; this was observed for inclinations up to 45°. The angle of faulting is strongly affected in the 60° and 75° orientations as well. Faults were unaffected by cleavage only for maximum compression perpendicular to cleavage.

The angle of faulting can thus be expected to vary considerably in anisotropic rocks; it is dependent on the inclination of the anisotropy

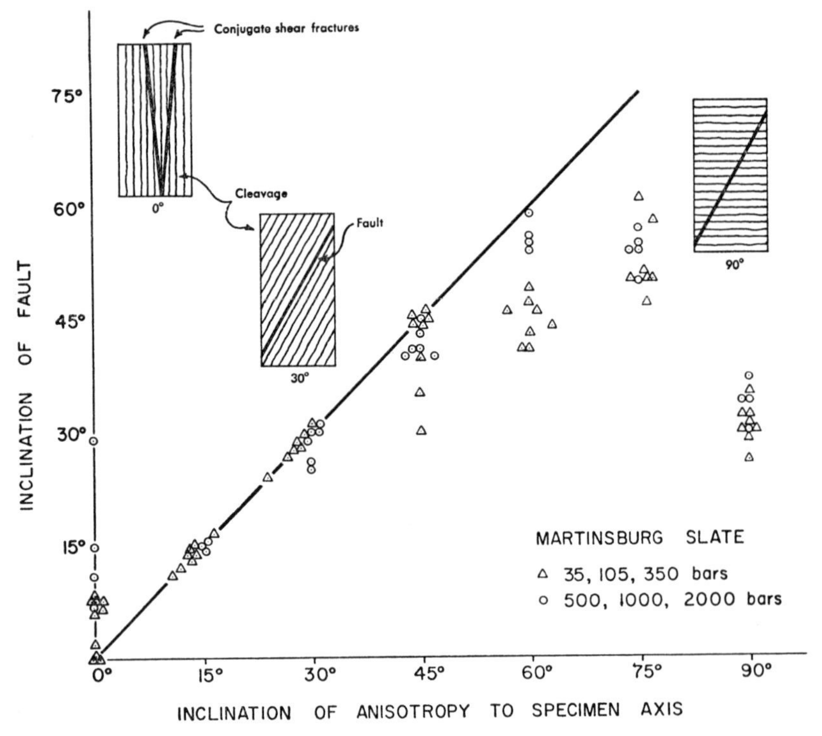

Fig. 3.—Effect of cleavage on the angle of faulting, Martinsburg slate. (Donath, 1963.)

with respect to the principal stress directions. In folding, as well, anisotropy plays an important part in determining the mechanism and, hence, the geometry of deformation (Donath and Parker, in press). Proper evaluation of the role of anisotropy in geologic deformation is therefore also of prime importance, for it is critical to fundamental problems treating the relation between structure and stress distribution.

MECHANISMS OF FOLDING AS FUNCTIONS OF DUCTILITY AND ANISOTROPY

For a further illustration of the importance of ductility and anisotropy in geologic deformation, consider their influence in folding. Plate II shows an overturned fold in limestone. An immediately obvious char-

PLATE I

Appearance of Wombeyan marble after testing.

Upper, After deformation at lower confining pressures. From left to right: (*a*) atmospheric pressure; (*b*) 35 bars, 1 per cent strain; (*c*) 100 bars, 2 per cent strain; (*d*) 210 bars, 12.5 per cent strain.

Lower, After 20 per cent strain at higher confining pressures. From left to right: (*a*) undeformed; (*b*) 280 bars; (*c*) 460 bars; (*d*) 1,000 bars. (Paterson, 1958.)

PLATE II

Overturned fold in Beekmantown limestone, Pennsylvania

PLATE III

Syncline in Helderberg limestone, West Virginia. (Photo by F. B. Conger.)

PLATE IV

Fold in Barton River slate, Vermont. (Photo by C. G. Doll.)

PLATE V

Photomicrograph of transverse section through gastropod (*Goniobasis*). Section is parallel to bedding. Plane polarized light. (Friedman and Conger.)

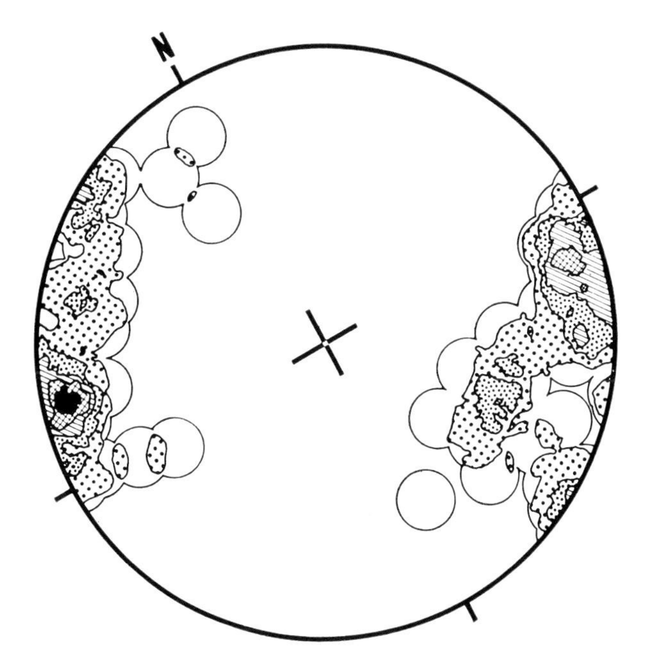

Equal-area lower hemisphere diagram illustrating the distribution of 70 compression axes derived from *e* twin lamellae in calcite grains of gastropod shell. (Friedman and Conger.) Note that the diagram is rotated to correspond to the orientation of the photograph.

acteristic of this fold is the uniform thickness of individual layers as measured perpendicular to bedding. Flow must have been negligible within individual layers or the thickness could not have remained constant. As illustrated, limestone can be very ductile and can flow readily even under low confining pressure and temperature. One can infer that folding must have occurred under conditions for which the rock had very low ductility. Folding was produced by slip between layers, and the layering, or anisotropy, was obviously important. The presence of anisotropy is, in fact, essential to flexural-slip folding.

A fold in another limestone sequence is shown in Plate III. In this fold the thickness of individual layers is not constant as measured perpendicular to bedding; many layers are thicker at the hinge than on the limbs. The limestone apparently was moderately ductile during deformation, and flow has occurred to differing degrees within the layers. The layering is still important, however, and controls the geometry of

TABLE 1

MECHANISMS OF FOLDING (DONATH AND PARKER, 1961)

Class	Type	Predominant Mechanism
Flexural.......	Flexural slip	Slip between flexed layers
	Flexural flow	Flow within flexed layers
Passive........	Passive flow	Flow across layer boundaries
	Passive slip	Slip across layer boundaries
Quasi-flexural..	....	Irregular flow within and across layers

deformation. Donath and Parker (1963) call this type of folding "flexural flow."

With increasing ductility, layering or other anisotropy becomes less and less important, and the involved rock tends to become mechanically isotropic. Under these conditions, layering is passive and exercises little control on the deformation. A passive fold in slate is shown in Plate IV. In point of fact, the original stratification in many slates is inherently of minor physical consequence because it is frequently little more than color change related to differences in oxidation. Passive folding can therefore occur either because of the ineffectiveness of anisotropy owing to high ductility or because of the absence of effective anisotropy (Donath, 1962*b*). As seen in Plate IV, the thickness of individual layers is, as in flexural-flow folds, not constant as measured perpendicular to bedding; it is nearly constant as measured parallel to the axial surface, however, which is a characteristic of ideal passive folds. The basic difference between a passive-flow fold and a flexural-flow fold is that, in passive-flow folding, flow occurs across layer boundaries, essentially parallel to the axial surface, but, in flexural-flow folding, flow occurs within the layers.

Where flow is irregular, certain layers may undergo flexural or pseudo-flexural folding at the same time that associated rocks show passive behavior. Ptygmatic folds and many folds showing disharmonic relations are examples of this gradational type of folding, which is here called quasi-flexural. The mechanisms of folding are summarized in table 1.

The type of folding that occurs is clearly a function of the ductility and the nature of inherent anisotropy of the involved rocks. This relationship can be expressed in terms of the mean ductility of the rock sequence and the contrast of ductilities within the sequence, as illustrated in figure 4. A rock sequence with a low mean ductility will fold by flexural slip; one with high mean ductility will undergo passive folding. If ductility contrast is high, however, passive folding may not be possible even at high mean ductility. For a layered sequence having little ductility contrast, such as a limestone sequence, the type of folding that will prevail is simply a function of the ductility of the involved rocks under the conditions of deformation. Point *A* in figure 4 could represent such a sequence deformed under near-surface conditions. Un-

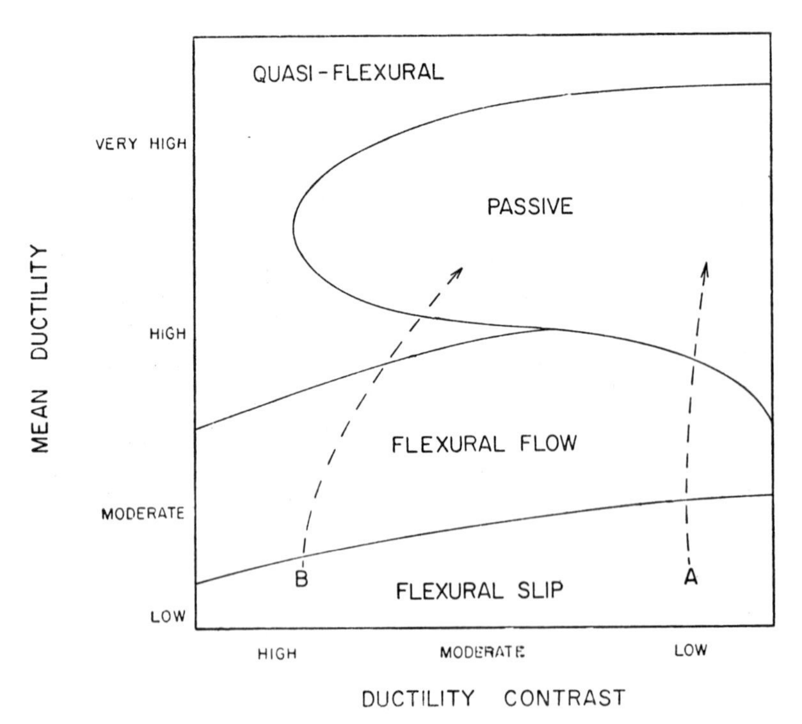

Fig. 4.—Fields of folding related to mean ductility and ductility contrast

der these conditions the mean ductility is low, and folding is by flexural slip. If the rock were deformed at greater depth and hence under higher confining pressure and higher temperature, the mean ductility would be raised, and folding would be either flexural flow or passive. A limestone-shale sequence may show high ductility contrast under near-surface conditions, as represented by point *B* in figure 4. Folding under these conditions would again be by flexural slip. With higher confining pressure or higher temperature, the mean ductility would be raised and the ductility contrast lowered. Depending on the specific conditions at the time of deformation, the folding mechanism could be flexural flow, quasi-flexural, or passive.

Thus, in folding, as well as in faulting, ductility and anisotropy determine the mechanisms that operate and the geometry of deformation. Recognizing the importance of these two factors in geologic deformation, how can one use this knowledge to learn more about the relation between a structure and its deformational environment? A qualitative evaluation of the role of ductility and anisotropy may very likely answer the basic question "Why has this particular structure developed?" and perhaps may even provide information about certain conditions of deformation, such as the approximate amount of overburden. But the answers to the questions "Under what physical conditions did the structure form?" and "How is the structure related in space and time to the stress responsible for it?" must stem from determinations in the structure of stress or strain distributions and ductility.

Methods of Determining Strain or Stress Distribution and Ductility

Determination of Ductility in Folds

Several methods based on simple assumptions regarding the geometry and homogeneity of flow can be used to determine ductility in folds. If certain of these assumptions do not hold for parts of a given structure, the values obtained represent only approximations; but the simplicity of technique places the methods at the disposal of all field geologists, and the information obtained would add immensely to our understanding of folding.

The permanent strain that a rock has undergone is a measure of ductility, provided this strain does not include displacement along a fault or fracture. Reasonable estimates of permanent strain, and hence ductility, can be obtained for individual layers in a flexural-slip fold from a simple relation between the layer thickness and its radius of curvature. Both can readily be measured in the field or determined from photographs. If it is assumed that the strain is negligible along the lower surface of a layer folded by flexural slip, the maximum strain in the layer will be along its upper surface and will be equal to the ratio of thickness of the layer to the radius of curvature of its lower surface. The relationship is shown in figure 5 along with a plot of strain ($\in$) *versus* the radius to thickness ratio (r/t). Strain here represents an integrated value for the upper surface of a given layer. A critical range of values for the radius-to-thickness ratio exists, approximately 1 to 10,

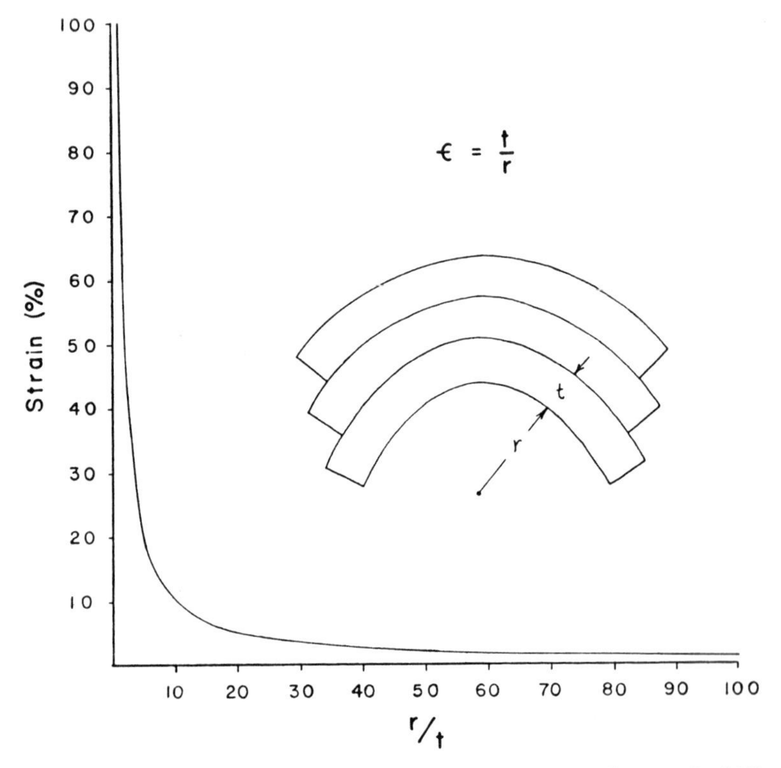

FIG. 5.—Effect of radius to thickness ratio (r/t) on strain ($\in$) in flexural slip folding

in which a small change in the ratio produces a significant change in the strain. For a ratio of radius to thickness of 20 or larger, the maximum strain is less than 5 per cent. This explains why thin-bedded sequences can undergo moderately intense folding by flexural slip with negligible strain in individual beds. Permanent strain estimated in this manner would represent a maximum value of ductility for the involved rocks because it is assumed that the "neutral axis" (points of no strain) coincides with the base of the layer and because fracture may have occurred in the layers before folding was completed.

For passive folds initiated by flexural slip, one can apply the method of analysis described by Ramsay (1962). In figure 6, *B*, x is the amount of strain perpendicular to the axial surface produced by passive folding. For a given layer, the ratio of thickness on a limb of the fold, as measured between tangents having the same dip on upper and lower surfaces, to that at the hinge is a function of the flattening, x. As shown in figure 7, the ratio of measured thicknesses, t'', can be plotted against the

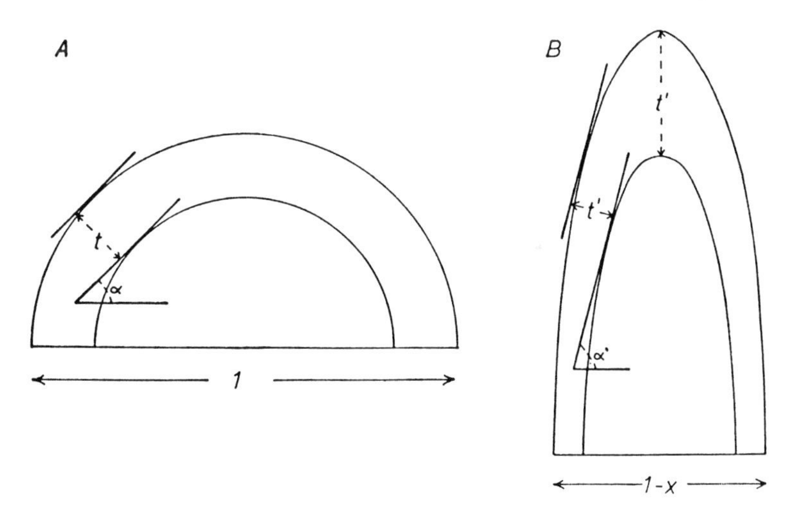

FIG. 6.—Passive deformation of a fold initiated by flexural slip. *A*, Bed of thickness t folded by flexural slip. *B*, Geometry of fold in *A* flattened by amount x; dip at corresponding points increases from α to α', thickness changes from t to t'. (Ramsay, 1962.)

respective angles of dip, α', and graphed. The flattening, or per cent strain normal to the axial surface, can then be read for measured values from natural folds. Strain thus determined does not include that produced in the rocks during the stage of flexural-slip folding.

DIRECT-MEASUREMENT TECHNIQUES

More sophisticated techniques can be used to determine the principal strains or the directions of causative principal stresses. The determination of stress or strain distributions in field structures is the essential link between field observation and theory. The variation of strain or of stress intensity or direction throughout a structure may conform to that predicted by one or another rheologic theory. At the least, such determinations would allow us to discard theories obviously not in accord with measurements. These direct-measurement techniques can be grouped into three principal categories: measurement of deformed objects within the rocks, determination of principal stress directions as based on known relations between stress and deformational feature, and measurement of elastic strain recovery.

Measurement of deformed oölites, pebbles, or fossils can provide information about both the magnitudes and the directions of principal strains, though the amount of strain indicated by a deformed object may not represent the *total* strain in the matrix if there is great disparity in

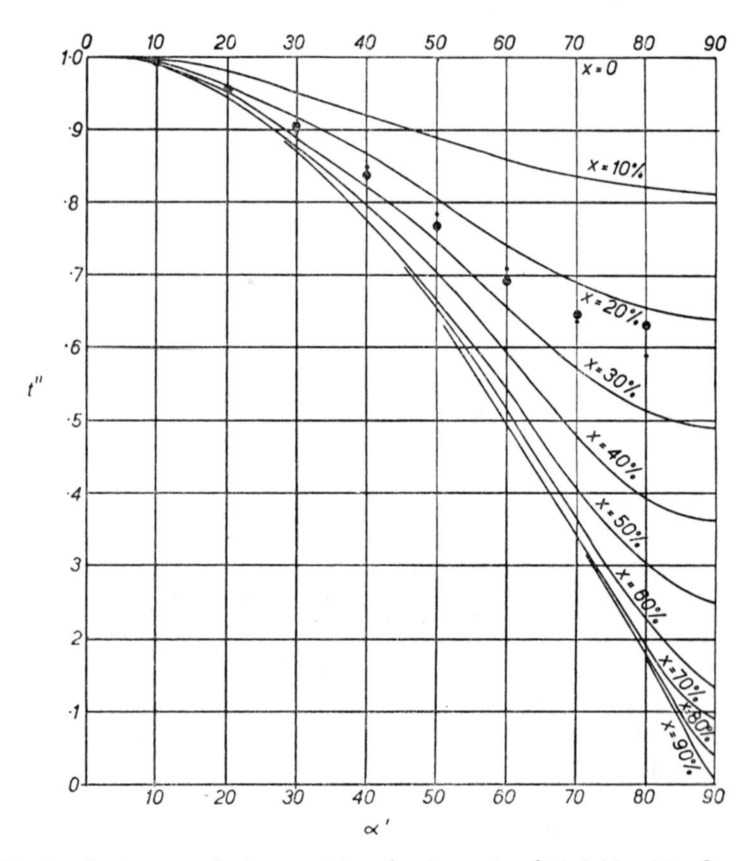

FIG. 7.—Strain perpendicular to axial surface in passive folds initiated by flexural slip, x, determined from the relationship between ratio of changed thicknesses (t'') to corresponding dip angles (α'). ($t'' = t'$ on limb/t' at hinge.) (Ramsay, 1962.)

the ductilities of the respective materials. Unfortunately, very few studies of this type have been made (cf. Cloos, 1947), even though a study of naturally deformed objects is one of the more obvious and most promising means of relating theory and experiment to field observations. An originally spherical object deformed to an ellipsoid can clearly provide the directions and magnitudes of principal strains. However, as pointed out by Brace (1960, 1961), the state of strain in a given deformation plane can be as readily determined from the relations between two originally perpendicular lines of known length ratio or from any three lines whose original angular relations are known. Thus, deformed fossils, bedding and cross-bedding relations, and similar features may all prove useful in strain analysis.

Even when it is not possible to define completely the state of strain, one may be able to define the direction of maximum elongation. This can be valuable, for example, in determining whether passive or flexural folding has occurred or whether the rocks have been affected by more than one deformation. Maximum elongation in a passive fold would tend

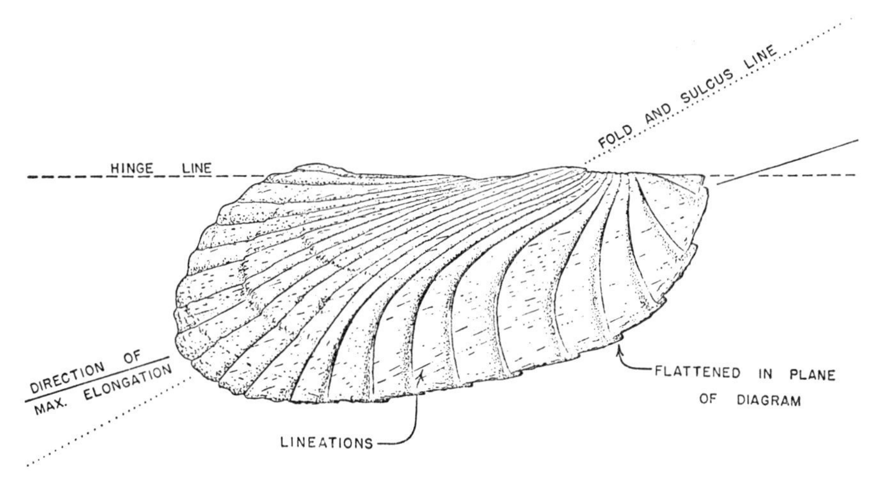

FIG. 8.—Deformed brachipod (*Dinorthis*), Middlebury synclinorium, Vermont (Crosby, Ph.D. thesis, Columbia University.)

to lie in surfaces parallel to the axial surface. Maximum elongation in a flexural-flow would ideally be perpendicular to the fold axis and would lie in surfaces not parallel to the axial surface.

Figure 8 is a sketch of a typical deformed brachiopod from the Middlebury synclinorium of Vermont. The fossil lies in the plane of the diagram, and its hinge line and sulcus line were initially perpendicular; they now are inclined toward one another at an acute angle, within which lies the direction of maximum elongation. Plications on the shell parallel to the direction of maximum elongation remain symmetrical, whereas those originally transverse to this direction are highly asymmetrical and flattened in the plane of a prominent cleavage. Lineations on the cleavage surfaces are parallel to the direction of maximum elongation. Crosby (Ph.D. thesis, Columbia University) has used fossil elongations to help differentiate the effects of different periods of folding in the Middlebury synclinorium by relating the elongations to specific fold mechanisms.

The second major category of direct-measurement techniques is concerned with the determination of principal stress directions as based on known relations between stress and a deformational feature. Fractures are very useful in this respect because the relationship between fracture and stress distribution has been well established. Two types of fractures are recognized (fig. 9): *extension fractures*, characterized by initial dis-

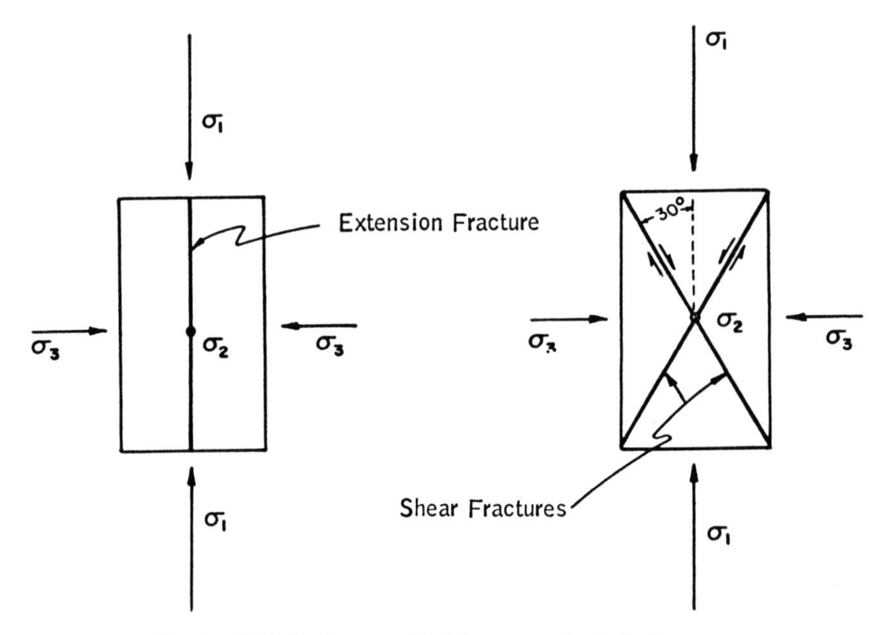

FIG. 9.—Relation between ideal fracture and principal stresses

placement perpendicular to the fracture surface and oriented perpendicular to the direction of least compression, and *shear fractures*, characterized by initial displacement parallel to the fracture surface and oriented at approximately 30° to the direction of maximum compression (Griggs and Handin, 1960). In theory, either or both of two shear fractures representing a conjugate pair could develop, having an included angle of approximately 60° that is bisected by the direction of maximum compression. Both types of fracture develop parallel to the direction of intermediate principal stress, and the intersection of any two fractures *formed in the same stress system* defines this direction. For proved conjugate shears, the causative stress directions can be determined with certainty. For a single shear fracture, the direction and sense of displacement must be known to determine the causative principal stress directions—though, if the shear is parallel to anisotropy, even this information is not sufficient (see Donath, 1961). For a single extension fracture, only the direction of least compression can be determined without additional information.

Fractures in a folded sequence of rocks can, for example, provide information about the attitudes of principal stresses in different parts of the fold. Figure 10 illustrates diagrammatically several common orientations of conjugate shear fractures in folds. The orientations at *B*, *C*, and *E* indicate a change in direction of local maximum compression from parallel to the fold axis at *B*, perpendicular to bedding at *C*, to parallel to bedding and perpendicular to the fold axis at *E*. The fractures at *A* and *D* are perpendicular to bedding and indicate local maximum

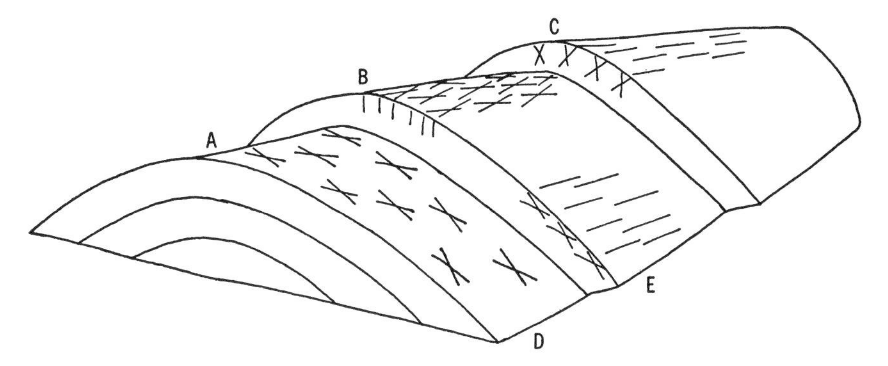

Fig. 10.—Common orientations of conjugate shear fractures in a flexural fold

compression parallel to bedding and perpendicular to the fold axis. Though it is unlikely that orientations such as those illustrated would be observed in close proximity to one another, each could occur at different places in the same structure.

Different orientations of local maximum compression in a fold generally are found in layers of low ductility and hence are most common in flexural-slip folds. In the example illustrated in figure 10, bending of the intermediate layer has caused extension parallel to the layer and perpendicular to the fold axis at *B;* this direction thus became the direction of greatest relief or least compression. The stress parallel to the fold axis was essentially unaffected and, being greater than the vertical stress in this example, became the maximum principal stress. Similarly at *C*, extension in the upper part of the layer controlled the direction of least compression; here, however, the maximum compression was perpendicular to the layer. At *A*, *D*, and *E* the maximum compression showed the same relation to layering and to the fold axis, but the direction of greatest relief (least compression) was perpendicular to the layering at *E*, whereas it was parallel to the fold axis at *A* and *D*. The orientations at *A* and *D* could have resulted from a slight extension parallel to the fold axis, possibly reflecting a culmination. If, however, everywhere in the fold the fractures show a constant orientation with respect to bedding, such as at *A* and *D*, one might question whether the fractures are related to the folding process or whether they were present before folding.

The type of analysis described above is a potentially valuable technique for determining the stress distribution in folded structure. To the writer's knowledge, no study of this type has been published, although the stress distribution responsible for a complex fault system in south-central Oregon has been determined in this manner (Donath, 1962*a*). Unfortunately, many fracture studies have yielded indecisive results or erroneous conclusions because the investigator has grouped fractures of all degrees of development from too large an area, has failed to distinguish among fractures of different ages or between extension and shear fractures, or has ignored the possible effects of anisotropy—to mention several of the more common pitfalls. The indiscriminate ("unbiased") lumping of all fracture data, followed by purely geometrical analysis, commonly leads to chaos, for the basically simple relations between fractures and stress distribution are often complicated by local changes in stress orientation or superposition of fracture systems.

The relations so well established for macroscopic fractures and stress orientation have been demonstrated to be equally valid on a microscopic scale. From results of experimentally deformed cylinders of calcite-cemented sandstones, Friedman (1963) concluded that, statistically, both the calcite cement and the detrital grains deform in response to principal stresses across the boundaries of the specimen rather than to local stress concentrations at grain contacts. Accordingly, extension fractures in the detrital grains are oriented perpendicular to the direction of least compression, and shear fractures are inclined at approximately 30° to the direction of maximum compression acting at the *boundaries of the specimen*. Thin-section study of microfractures in oriented specimens collected from different parts of a structure thus also represents a potentially valuable method of mapping the stress distribution in the structure.

Another important technique in this second category is the determination of principal stress directions based on crystal deformation. From studies of experimentally deformed calcite single crystals and marble, Turner (1953) presented a dynamic interpretation of calcite twin lamellae, pointing out that unique directions of maximum and minimum compression can be located for which the development of observed twin lamellae is most favored. Glide mechanisms in both calcite and dolomite have been used in studies of naturally deformed rocks to determine compression and extension axes (cf. McIntyre and Turner, 1953; Weiss, 1954; Crampton, 1958; Christie, 1958).

Friedman and Conger (in press) have recently applied the technique to a naturally deformed gastropod shell in an oriented specimen of limestone collected from a syncline (Pl. V). The transverse section through the gastropod is parallel to bedding. Measured *c*-axes of calcite lie at high angles to the wall of the shell and thus form a peripheral girdle when plotted on an equal-area net. Calcite crystals along two opposite sides of the shell (each encompassing about 120° of arc) are profusely twinned; those along the two remaining 60° arcs are poorly twinned. Compression and extension axes derived from *c*-axes and twin lamellae are strongly oriented and are in good agreement with the field structure. The lower part of Plate V shows the distribution of compression axes as determined from 70 grains. The center of these concentrations marks the derived position of maximum compression that lies about 15° from normal to the fold axis.

The third major category of direct-measurement techniques makes use of the fact that stored elastic-strain energy is present in many rocks. The strain recovered by relieving the rocks can be measured, and from this the directions and magnitudes of the residual principal stresses can be determined. Several techniques of measurement have been employed, but all operate on essentially the same principle: a strain indicator is affixed to the rock surface or placed in a hole in the rock, then a stress-release channel is cut by overcoring the strain indicator. Thus the rock is allowed to expand and the elastic-strain recovery can be measured (Hast, 1958).

Preliminary work suggests that a stress fabric can be imprinted in rocks much as a metamorphic fabric is in highly deformed rocks. If this proves to be the general case, some very exciting information can be obtained about stress distributions in geologic structures, and considerable light may be shed on the problem of local reorientation of stress adjacent to faults or within folds. The magnitudes of residual stresses would be expected to differ from those responsible for the structure to a degree that is a function of postdeformational time and history. Nevertheless, the directions of residual principal stresses would presumably coincide with those of the deforming stresses and could be related to the deformational geometry.

Theoretical and Model Studies

Ideally, the structural geologist would like to formulate the relationship between the deformational environment and the resultant structure. Ductility measurements in geologic structures, interpreted in the light of experimental work, and determination of strain or stress distributions would contribute immensely to the possibility of such formulation. If a relationship can be established among certain variables and the geologic effects, reasonable inferences about unexposed parts of a structure (e.g., relations at depth) and about the physical and chemical environment that obtained during its development can be made from limited information.

In an effort to predict the characteristics of unexposed parts of a structure, to explain associated structure, or to discover new, previously unsuspected features, geologists have on occasion appealed to both theoretical and model studies. Folding can, for example, be treated theoretically if the material is assumed to be ideally elastic, plastic, or viscoelastic. Relationships among wave length, amplitude, and other basic fold characteristics can be determined; moreover, the relations between boundary conditions and internal stress distribution can be established. But what if the initial assumption is basically incorrect—if the rocks are not perfectly elastic, plastic, or viscoelastic? Are the conclusions based on the analysis then correct? A valid rheologic model cannot be selected until more is known about the properties of geologic materials and about stress distributions actually associated with structures. At our present state of knowledge it may be a bit premature, and could be misleading, to apply certain theory to actual field relations. Enough is known about the behavior of rocks to show that no simple rheologic model will explain all aspects of rock behavior. Nevertheless, the proper evaluation of data from experimental work and from even rather simple field observations, several of which have been discussed in this paper, may allow one to choose a reasonable rheologic model.

Can model studies provide the answers? Certainly, anyone who has worked with or has observed scale models must have been impressed by the clear and obvious relations among structures, their sequence of development, and their relations to the boundary forces that produced them. It should be equally clear, however, that the model must be properly scaled (Hubbert, 1937) and that, if the material itself differs from the original or prototype, certain mechanisms of deformation must necessarily differ.

Unfortunately, it is not possible to scale correctly most geologic structures. Body forces cannot be simulated in models, and, inasmuch as these can be extremely important in natural deformation, the model can be seriously at fault. Strength and ductility, which are independent variables, cannot be correctly scaled, nor can the increase in confining pressure with depth. The small variation allowable in certain properties also presents severe restrictions. Density can be varied at most by a factor of, say, two; whereas scaling of the natural environment may require factors of a million. In many cases the properties of the prototype are not even known and the properties of the model must be selected arbitrarily. Moreover, boundary conditions of the structure in nature are unknown and therefore cannot be simulated in the model.

The writer concludes from such considerations that few important problems in dynamic structural geology will be solved from model studies. This is not to say that model studies are not important, however, for demonstration purposes or as analogs for the solution of specific problems in theory. Many theoretical treatments are too unwieldy to be handled easily by mathematical analysis. Model studies, such as photoelastic techniques, may provide easily grasped solutions. The properties of the model must, however, comply with the theory for the analog to be valid.

In two excellent papers recently published (Hubbert, 1951; Sanford, 1959) the problem of relating faulting to boundary conditions was treated both theoretically and with model studies. The mathematical analysis in each case was based on elasticity theory; the model studies utilized sand as the model material. Sand, not being a perfectly elastic medium, does not satisfy the requirements of elasticity theory. Yet, in the work cited, the results obtained from the model studies were consistent with those predicted by the theoretical treatment. One can conclude from such results that a given model may not demonstrate a *unique* theory. The implications of this should be clear: though certain model (or field) relations are in accord with those predicted by a particular theory, this does not in itself prove the theory, and erroneous conclusions could be drawn from the extended application of the theory.

Equally important is the fact that even if a model material does comply with a particular rheologic theory and certain obtained relations do greatly resemble those observed in the field, the mechanisms that operated and the boundary forces responsible for the gross relations could have been quite different in the model and prototype. An ever present danger in model studies lies in the possibility that the investigator will equate mechanisms in the model with those of geologic phenomena of an entirely different scale or material. In clay model studies, for example, the orientation and development of slip surfaces in the clay have been extended directly to fracture and fault relations in rocks, despite the obvious discrepancies in scaled properties, especially ductility and anisotropy. The very fact that the manner of preparation of the model material affects the results ought to give pause for thought.

Conclusions

The solution to most fundamental problems in dynamic structural geology requires an integrated approach involving field work, theory, and experiment. Experimental studies have emphasized the role of ductility and anisotropy in determining the mode and geometry of deformation, and field evidence supports the conclusions. Nevertheless, field behavior cannot yet be translated into environmental conditions responsible for the behavior; ductility measurements must first be made in geologic structures. Theoretical treatment must wait until stress and strain distributions are determined in field structures and the behavior of geologic materials is better known, for, until this is done, one or another rheologic theory cannot be considered most reasonable. Because models of geologic structures cannot be dynamically scaled, model studies are valuable primarily as analogs for particular theories or for demonstration.

It seems, therefore, that the most important work in dynamic structural geology in the immediate future will be the application of existing or new techniques to determine ductility, strain, and stress distributions in natural structures and the interpretation of these structures in the light of this acquired knowledge.

References Cited

Brace, W. F., 1960, Analysis of large two dimensional strain in deformed rocks: Internat. Geol. Cong., 21st, Norden, pt. 18, p. 261–269.

——— 1961, Mohr construction in the analysis of large geologic strain: Geol. Soc. America Bull., v. 72, p. 1059–1080.

Christie, J. M., 1958, Dynamic interpretation of the fabric of a dolomite from the Moine thrust-zone in north-west Scotland: Am. Jour. Sci., v. 256, p. 159–170.

Cloos, E., 1947, Oolite deformation in the South Mountain Fold, Maryland: Geol. Soc. America Bull., v. 58, p. 843–918.

Crampton, C. B., 1958, Structural petrology of Cambro-Ordovician limestones of the northwest Highlands of Scotland: Am. Jour. Sci., v. 256, p. 145–158.

Crosby, G.W., 1963, Structural evolution of the Middlebury synclinorium, west-central Vermont: Ph.D. thesis, Columbia University.

Donath, F. A., 1961, Experimental study of shear failure in anisotropic rocks: Geol. Soc. America Bull., v. 72, p. 985–990.

——— 1962*a*, Analysis of Basin-Range structure, south-central Oregon: *ibid.*, v. 73, p. 1–16.

——— 1962*b*, Role of layering in geologic deformation: N.Y. Acad. Sci. Trans., v. 24, p. 236–249.

——— 1963, Strength variation and deformational behavior in anisotropic rock: Proc. Internat. Conf. State of Stress in Earth's Crust, RAND Corporation.

——— and Parker, R. B., in press, Folds and folding: Geol. Soc. America Bull. (Geol. Abs. for 1961. Geol. Soc. America Spec. Paper 68, p. 87).

Friedman, M., 1963, Petrofabric analysis of experimentally deformed calcite-cemented sandstones: Jour. Geology, v. 71, p. 12–37.

——— and Conger, F. B., in press, Dynamic interpretation of calcite twin lamellae in a naturally deformed fossil: *ibid.*, v. 72.

Griggs, D., and Handin, J., 1960, Observations on fracture and a hypothesis of earthquakes: Geol. Soc. America Mem. 79, p. 347–373.

Handin, J., and Hager, R. V., Jr., 1957, Experimental deformation of sedimentary rocks under confining pressure: tests at room temperature on dry samples: Am. Assoc. Petroleum Geologists Bull., v. 41, p. 1–50.

——— ——— 1958, Experimental deformation of sedimentary rocks under confining pressure: tests at high temperature: *ibid.*, v. 42, p. 2892–2934.

Hast, N., 1958, The measurement of rock pressure in mines: Stockholm, Sveriges Geologiska Undersokning Arsbok 51, no. 3.

Heard, H. C., 1960, Transition from brittle fracture to ductile flow in Solenhofen limestone as a function of temperature, confining pressure, and interstitial fluid pressure: Geol. Soc. America Mem. 79, p. 193–226.

——— 1963, Effect of large changes in strain rate in the experimental deformation of Yule Marble: Jour. Geology, v. 71, p. 162–195.

Hubbert, M. K., 1937, Theory of scale models as applied to the study of geologic structures: Geol. Soc. America Bull., v. 48, p. 1459–1520.

——— 1951, Mechanical basis for certain familiar geologic structures: *ibid.*, v. 62, p. 355–372.

McIntyre, D. B., and Turner, F. J., 1953, Petrofabric analysis of marbles from Mid-Strathspey and Strathavon: Geol. Mag., v. 90, p. 225–240.

Paterson, M. S., 1958, Experimental deformation and faulting in Wombeyan marble: Geol. Soc. America Bull., v. 69, p. 465–475.

Ramsay, J. G., 1962, The geometry and mechanics of formation of "similar" type folds: Jour. Geology, v. 70, p. 309–327.

Sanford, A. R., 1959, Analytical and experimental study of simple geologic structures: Geol. Soc. America Bull., v. 70, p. 19–52.

Serdengecti, S., and Boozer, G. D., 1961, The effects of strain rate and temperature on the behavior of rocks subjected to triaxial compression: 4th Symposium Rock Mechanics Proc., Bull. Mineral Industries, Nov. 1961, Penn. State University.

Turner, F. J., 1953, Nature and dynamic interpretation of deformation lamellae in calcite of three marbles: Am. Jour. Sci., v. 251, p. 276–298.

Weiss, L. E., 1954, A study of tectonic style: structural investigation of a marble-quartzite complex in Southern California: Univ. Calif. Pub. Geol. Sci., v. 30, p. 1–102.

S. S. WILKS

Statistical Inference in Geology

WHEN I WAS INVITED to participate in this symposium, it was explained that what was desired was a discussion by a statistician of some basic statistical philosophy and concepts that might be of interest to geologists in thinking about some of their problems, and not a presentation of details of various techniques for dealing with specific problems in geology. For one who is equally at home in geology and statistics that would be a fascinating challenge. But for a statistician like me who knows virtually no geology it is a bewildering assignment, which can be undertaken only with trepidation. So you can understand that I was hesitant about accepting the invitation. Once having accepted, however, and having browsed through a great deal of recent geological literature to get an over-all view of where and how methods of statistical analysis and inference have been introduced into geology and are currently being used there, I found the assignment to be very interesting. It is my impression from geological literature that significantly increasing numbers of applications of statistical methods have been made to problems in geology during the last few years. Furthermore, it seems to me that in many cases these methods are being very effectively applied, and there are various possibilities for further use of the methods.

In presenting my thoughts on statistical analysis and inference, with particular reference to problems in geology, I shall discuss in the following three sections some of the basic concepts underlying all statistical analysis and inference, such as precision and accuracy of measurement processes, statistical populations, sampling, randomization, distributions of measurements, estimation of population characteristics from samples,

Dr. Wilks, a native Texan, attended North Texas State College, the University of Texas, and the State University of Iowa, where he received the Ph.D. degree in 1931. He has been a member of the Department of Mathematics of Princeton University since 1936 and has been professor of mathematical statistics since 1944. He has been very active in scientific and governmental advisory bodies and is the author of numerous papers and several books on mathematical statistics.

This paper was prepared with the partial support of an Office of Naval Research Contract.

and testing hypotheses about populations from information contained in samples. In the remaining sections of the paper I shall discuss, without going into technical details, the basic ideas and objectives of certain statistical techniques that seem to be particularly relevant to geological studies, namely, linear regression analysis, analyses of variance, and discriminant analysis. Such details can be found in various statistics books, which will be referred to as we proceed. No attempt has been made to discuss all statistical techniques that have been or are being used in geological studies. For example, methods for dealing with orientation data, time series, and other more highly specialized techniques particularly applicable to geophysical problems have not been included. The reader is referred particularly to the book by Miller and Kahn (1962) for details of the various statistical methods as they apply to problems in the geological sciences.

Precision and Accuracy of a Measurement Process

Any scientific study of an object involves the application of well-defined observational or measurement processes to that object. The process can be very simple, like determining weight or length of an object, or it can be much more complicated, like determining weight per cent of silica (SiO_2) or titania (TiO_2) in a rock specimen, or determining the age of a skeleton by the carbon-14 method. If the measurement process is to have scientific validity, the application of the process to the object must be repeatable "under essentially similar conditions," with the result that outcomes of repeated applications of the measurement process are very similar or cluster together in some measurable sense. The extent to which such outcomes cluster together constitutes the notion of *precision* of the measurement process—the closer the clustering of the repeated outcomes, the higher the precision. (In some fields of science the word *reliability* is used rather than *precision*.) There are various statistical techniques for measuring precision of a measurement process, the most common ones being the standard deviation of a set of repeated measurements, or the coefficient of variation (ratio of standard deviation to mean) of the set of measurements produced by repeated applications of the measurement process. Thus, if $x_1, x_2, \ldots, x_n$ is such a set of measurements, the mean $\bar{x}$ is defined as

$$\frac{x_1 + x_2 + \ldots + x_n}{n},$$

and the standard deviation s as the square root of the quantity

$$\frac{(x_1 - \bar{x}_2)^2 + (x_2 - \bar{x})^2 + \ldots + (x_n - \bar{x})^2}{n - 1}.$$

In cases where it is sufficient merely to know something about precision of measurement relative to variability of the objects under study, a useful procedure is to study the pairs of measurements obtained by making a first and second measurement on each object in a random sample of the objects, using correlation, regression, and other two-dimensional statistical methods for analyzing the resulting pairs of measurements. In tracking down the main sources of variability in the measurement process, and to eliminate them or to reduce their effects and hence to increase the precision of a measurement process, fairly sophisticated statistical techniques may be required, such as design of experiment and analysis of variance techniques. An interesting example of such a technique is contained in a paper by Flanagan (1960) dealing with the problem of precision and accuracy of procedures for determining lead content of a specified granite widely known in geological circles as G-1. This example will be discussed further in the section "Analysis of Variance."

The high precision of a meaurement process is not sufficient for establishing the scientific validity of the process. The outcomes of repeated applications of the measurement process not only must have an adequate degree of precision but must also have *accuracy*. That is, they not only must cluster together but must cluster together around the "correct value" of the quantity that is supposedly being measured. It should be noted that we talk about precision and accuracy not of a measurement but of a measurement process. That is, we must think in terms of a series of measurements made by repeated applications of the process and not in terms of a single measurement. (In some fields of science the term *validity* is used rather than *accuracy*.)

The concept of accuracy is particularly important in a situation in which a new measurement process is devised that will be easier, faster, or cheaper to apply than a firmly established "standard method." In such a case it is clearly not sufficient to establish precision of the new method by repeated runs of it, but accuracy must be established by calibrating the new method against a "standard method" of known accuracy or against a series of specimens having various known values of the quantity to be measured by the new method. For instance, a colorimetric method is commonly used for determining alpha-resin content of beer. But this fast and cheap method had to be calibrated against samples of beer with known alpha-resin content as defined by "standard" chemical procedures. A good example involving such calibration in geology is the introduction of spectrochemical methods for making quick quantitative determinations of certain chemical constituents in rocks. These newer methods, which are discussed, for example, by Ahrens and Taylor (1961), have been calibrated against the older "wet chemistry" methods as standards, at least for constituents with higher concentrations, such as silica (SiO_2), alumina (Al_2O_3), lime (CaO), potash (K_2O), etc., the calibration for any given constituent being accomplished by a "working curve" showing the relationship between spectrographic readings and known concentrations of the constituent. An interesting study of precision and accuracy in quantitative analysis of major elements in granitic rocks by X-ray fluorescence has been made by Baird, MacColl, and McIntyre (1961).

It is a truism to say that a measurement process, in order to have scientific validity, must possess reasonable degrees of precision and accuracy. But in a paper of this sort it should be explicitly stated. There is all too often a tendency in scientific investigations to assume that a measurement process has these properties without actually going through the procedure of establishing the degree of precision or of accuracy of the process. The minimum requirement for obtaining information on precision is to make duplicate runs of a measurement process. In the case of accuracy, the measurement process must be calibrated against some independently established "standard" measurement process. And even then satisfactory degrees of precision and accuracy may not be easily obtained. Perfect precision and accuracy are rarely, if ever, achieved in practice. The best that can be hoped for is a gradual refinement and control of the measurement process through successive experimentation to the point where precision and accuracy are high enough for practical purposes. An interesting example, familiar to geologists and illustrating the unsuspected difficulties in achieving accuracy, is a large collaborative study of the United States Geological Survey, undertaken largely through the initiative and under the guidance of H. W. Fairbairn of the Massachusetts Institute of Technology, in which samples of a granite designated as G-1 from a quarry near Westerly, Rhode Island, and samples of a diabase designated as W-1 from a quarry near Centerville, Virginia, were chemically analyzed by thirty-four collaborating chemists at laboratories throughout the world. Extensive statistical analysis of the results of this experiment has been published by Dennen, Ahrens, and Fairbairn (1951). Subsequently, over thirty new analyses of G-1 and W-1 were reported by various laboratories over the world and were statistically analyzed by Stevens and Niles (1960).

The quantitative determinations of the main constituents of G-1 and W-1 by standard chemical procedures as reported in the two studies are given in table 1, where n is the number, $\bar{x}$ is the mean, and s is the standard deviation of the determinations made in each case. Thus, s as an indicator of precision is a measure of the actual variability of the determinations around the value of $\bar{x}$ in any given case. The ratio $s/\bar{x}$

provides an indicator of precision as a measure of the variability of the determinations around $\bar{x}$ relative to $\bar{x}$ itself.

In data such as that given in table 1, a rough rule of thumb for interpreting n, $\bar{x}$, and s is that about 50 per cent of the n individual determinations of any given constituent in any given set of data (1951 or 1960) would lie inside the interval $\bar{x} \pm \frac{2}{3}s$, and about 68 per cent would lie inside the interval $\bar{x} \pm$ s. (For instance, about 17 of the 34 determinations of weight per cent Fe_2O_3 in G-1 from the 1951 data would lie inside of 0.92 $\pm$ 0.20 and about 22 inside of 0.92 $\pm$ 0.29. The actual numbers were 17 and 27.)

TABLE 1

INFORMATION ON THE PRECISION OF MEASURING MAIN CONSTITUENTS IN G-1 AND W-1 FOR 1951 AND 1960 DATA (DATA FROM STEVENS AND NILES, 1960)

CONSTITUENT	G-1						W-1					
	1951 Data			1960 Data			1951 Data			1960 Data		
	n	$\bar{x}$	s	n	$\bar{x}$	s	n	$\bar{x}$	s	n	$\bar{x}$	s
SiO_2	34	72.24	0.37	26	72.49	0.19	30	52.33	0.35	30	52.48	0.31
Al_2O_2	34	14.34	.46	26	14.30	.22	30	15.17	.85	30	15.05	.26
Fe_2O_3	34	.92	.29	23	.98	.32	30	1.75	.95	28	1.49	.29
FeO	34	1.01	.12	23	.96	.09	30	8.55	.69	28	8.72	.18
MgO+0.63 BaO	34	.39	.14	25	.40	.11	30	6.53	.47	29	6.62	.13
CaO+SrO	34	1.40	.14	25	1.39	.11	30	10.96	.17	29	10.98	.15
Na_2O	34	3.30	.26	25	3.32	.19	30	2.07	.21	28	2.07	.20
K_2O	34	5.48	.46	25	5.35	.29	30	.70	.17	28	.64	.04
H_2O-	28	.05	.03	20	.08	.07	28	.15	.07	26	.17	.06
H_2O+	30	.39	.22	21	.32	.09	30	.57	.24	25	.50	.16
TiO_2	34	.25	.06	26	.25	.01	30	1.04	.23	30	1.10	.15
P_2O_5	31	.11	.07	23	.09	.02	29	.14	.04	29	.16	.07
MnO	33	.03	.01	23	.03	.01	30	.18	.07	29	.16	.02

It is evident from examining the numbers in table 1 that the relative precision (ratio of s to $\bar{x}$) of measuring various constituents depends heavily on the abundance of the constituent, the precision in some cases being very low. There is a noticeable increase in the improvement of precision over the period from 1951 to 1960, owing largely, presumably, to the tightening-up of experimental procedures as a result of the shock produced by studying the laboratory-to-laboratory variation in results. It should be pointed out that intralaboratory precision studies of the various petrochemical methods used in analyzing G-1 and W-1 have been made for many years, but the two studies mentioned above and conducted by the United States Geological Survey seem to be the first extensive interlaboratory investigation of the precision of the methods.

The most significant finding of the two United States Geological

Survey studies was the surprisingly large interlaboratory variability of results, discovered in the 1951 study, and the considerable improvement in precision that occurred during the subsequent nine years, as reported in the 1960 study. Such interlaboratory variability resulting from repeated applications of standard measurement procedures is not unique to geochemistry. They have been made for many different measurement processes, usually with the finding of surprisingly large variations from laboratory to laboratory. Such studies emphasize the difficulties of establishing high degrees of precision and accuracy of measurement processes in practice and also the need for continuing effort to develop and maintain standardized procedures for measurement processes. Design of experiment and analysis of variance techniques, to be discussed in the section on analysis of variance, are useful for identifying sources of errors in measurement processes and measuring their magnitude, in any systematic effort to improve the precision and accuracy of such processes.

Populations and Samples

In many investigations the basic entity for study is a large collection of objects or individuals commonly called a *population* or *universe*. For example, in an investigation of quality of transistors produced in a factory the population of interest might be the transistors of a certain type produced during a given day. Or in a study of New Jersey automobiles the population might be taken as the list of all automobiles registered in New Jersey as of a given date. In a geologic study the population of interest might be the pebbles over a certain size in a specified deposit, or it might be the sand on a portion of a beach or a particular granite mass. In the case of the transistors, the population is a well-defined set of objects. In the case of the pebbles, the population can be made precise by defining the boundary of the deposits. In the case of the granite mass, the population of objects involved is not so well defined. One possibility would be to consider the population as being formed by particles into which the granite mass could conceivably be crushed by a given process. Another would be to think of the population as being the set of cubes, one inch on a side, into which the mass could conceivably be cut, according to some standard procedure for cutting.

In studying a population of objects, we are usually interested in some measurement (or set of measurements) that could be made on each object in the population. Thus the engineer might be interested in the resistances of the transistors for quality-control purposes; the budget director of the state of New Jersey might be interested in the weight classes of automobiles registered in New Jersey on January 1, 1963, for revenue purposes; the geologist might be interested in sizes of the particles into which the granite mass might be crushed or in the silica content of the cubes of granite into which the mass might be cut for classification purposes. But, to make order and sense out of such a collection of measurements, we must think in terms of something like the average or median or some other function of the measurements. For many purposes it is convenient to think in terms of the *cumulative frequency distribution* of the collection of measurements, that is, a table or graph essentially showing the fraction of measurements less than or equal to each possible value a measurement could have. We could compute the mean, the median, the quartiles, and various other useful characteristics of the collection of measurements from such a distribution if it were available. There are various other ways of organizing or arranging a sample of measurements for study and analysis. Some of these will be discussed in later sections.

In most practical situations populations have very large, often astronomically large, numbers of objects in them, and it is unfeasible, if not completely unthinkable, to make the measurement of interest on every object in the population. Hence, it would not be feasible or possible to determine the cumulative frequency distribution of the measurements for the entire population. We must therefore settle for a procedure for selecting a manageably small number or *sample* of objects from the population and actually making our measurements on the objects in the sample. The problem, then, is to select the sample in such a way that the cumulative frequency distribution of measurements or any other function of the measurements on the sample items in which we may be interested will provide a faithful approximation to that for the entire population. The only way we can do this and be able to make probability statements about how closely the sample results will approximate the corresponding results for the entire population is to select the sample by randomization procedures. Essentially, this means drawing the sample from the population so that each object in the population has the same chance (or at least a known chance) of being drawn into the sample, or performing a similar operation on each of several subpopulations into which the population might be divided. There are many methods of using randomization procedures for sampling, such as simple random sampling, stratified sampling, nested sampling, cluster sampling, etc. We cannot go into a detailed technical discussion of these methods here. Such discussion can be found in various statistics books, such as those by Bennett and Franklin (1954), Cochran (1953), and Cox (1958). About all we can do here is discuss the main ideas of one or two of the simpler methods of sampling with some reference to geology populations.

In practice, each method of sampling is implemented by the use of random numbers. In the case of simple random sampling, we can think of numbering the N objects in the population serially from 1 to N. If we wish a simple random sample of n objects from a such population, we draw from a table of random numbers n (different) numbers in the set 1, 2, . . . , N in such a way that each of the numbers 1, 2, . . . , N has the same probability of being drawn. It should be pointed out that it is not necessary, even in simple random sampling, to think of the objects in a population as being physically serially numbered. It is sufficient to think of them as being systematically arranged so that they could be located and numbered by some rule if desired. In such a situation *nested sampling* can be used, in which a block of a specified number of objects is selected from the populations, and from this block a sub-block of a given number of objects is selected, and so on until a subdivision is reached from which one or more objects are drawn. Such a process is repeated until a sample of desired size is drawn. For instance, if one wishes to draw a telephone number at random from the 1961–62 Manhattan telephone directly, one can draw one of the 1767 pages at random, then one of the four columns on that page at random, then one of the 130 lines in that column at random. If this results in the initial line of a name and address, it is the required randomly selected telephone number, otherwise the entire operation is repeated until the initial line of an address is found. Repetition of this procedure n times will produce a random sample of n names. Such a procedure clearly does not require serial numbering of all addresses in the telephone directory. But they are arranged in a systematic manner by name and address, so they could be numbered if desired.

In case of sampling the granite mass, we can do so by randomly drawing a 1-inch cube, say, once the granite mass that is considered to be the population under study is defined. This could be done by locating the granite mass as a three-dimensional body with specified boundaries in a three-dimensional coordinate system that, for simplicity, say, extends from 1 to N_1 inches in the x_1 direction (e.g., east), from 1 to N_2 inches in the x_2 direction (e.g., north), and from 1 to N_3 inches in the x_3 direction (e.g., vertically downward), so that the entire granite mass to be studied would lie in the positive octant of the $x_1x_2x_3$ coordinate system. One would then draw a random number from 1, 2, . . . , N_1, one from 1, 2, . . . , N_2, and one from 1, 2, . . . , N_3. If the three random numbers thus drawn were (n_1, n_2, n_3), then these three numbers would define a specific 1-inch cube of space in the $x_1x_2x_3$ coordinate system. If it is occupied by granite, this is the desired cubic inch; if it is occupied by empty space, water, or some other substance, the entire drawing operation is repeated. Repeating this operation enough times to obtain a sample of M specified 1-inch cubes of granite from the granite mass as defined would constitute a simple random sample of M 1-inch cubes from the mass. If the mass is very large, the units could be taken as feet or yards, or some larger unit, in which case we could choose one of such units at random and then pick a 1-inch segment at random from it. Needless to say, the problem of actually extracting a specified cubic inch of granite would require very accurate extracting procedures. In practice, however, such a degree of accuracy would not be necessary. Instead of a simple random sample, a *systematic* or *grid* sample could be drawn, for example, by dividing N_1 into m_1 equal intervals and taking as selected values of x_1 the mid-points of these intervals. We would similarly divide N_2 into m_2 equal intervals and select their mid-points as selected values of x_2 and N_3 into m_3 equal intervals and select their mid-points as selected values of x_3. We would thus have, for suitably chosen values of m_1, m_2, m_3, a total of $m_1 \times m_2 \times m_3$ selected points in the $x_1x_2x_3$ space, at each of which we would select a 1-inch cube of granite. If the various 1-inch cubes of granite drawn into the sample by either of these methods were pooled together as a single batch of material and subjected to particle-size, chemical, or modal analysis—or any other kind of geological measurement process—the results would provide estimates of parameters for the entire granite mass if such estimates were desired—the larger the number of individual 1-inch cubes in the sample, the closer would be the estimates to the actual population values of the parameters. Or, what is likely to be more useful, if the different 1-inch cubes that make up the entire sample were analyzed separately, they would provide information about the variation of the particular characteristic being measured over the granite mass. For this purpose, the systematic or grid sample is clearly preferable, since it would provide a more systematic coverage of the granite mass than would the simple random sample.

It should be noted that, in order to implement either of the two types of sampling mentioned above, it is important that the particular geological population (e.g., the granite mass) being studied be carefully defined in terms of its boundaries in the coordinate system adopted. Such a mass must be defined in terms not only of lateral extent but also of depth. If portions of such specifically defined rock masses are overlain with water or other material, including other kinds of rock material, that makes them inaccessible for sampling, then these portions of the granite mass must be excluded from the population being defined and sampled.

All that has been said about sampling populations consisting of solid masses also applies to sampling of populations or masses of clay, silt, sand, pebbles, and other aggregates. If the populations are considered as masses with well-defined boundaries, they can be subjected to simple random sampling or grid sampling, as indicated above. The unit amount of material to be taken at each sample point would depend on the type of material sampled. For instance, sampling of pebble deposits would presumably require larger units than would sand or soil.

In the preceding discussion we have considered the problem of sampling a three-dimensional mass. Similar remarks apply to sampling populations that are essentially one-dimensional (e.g., a thin outcrop band) or two-dimensional (e.g. a layer of shale underlying a given region). In the one-dimensional case we could think of the x_1-axis as measuring distance along the band, the total distance being N_1 inches, say (or feet or yards). Hence, if we wanted to take a single section of material, we would draw a number at random from 1, 2, . . . , N_1 and take a section of material of a uniform width across the band centered at the number drawn. This process would be repeated n times if we wanted a simple random sample of n sections from the band. Or we could draw a systematic sample of m_1 sections by dividing N_1 into m_1 equal intervals and taking a section of material at the mid-point of each interval. For the two-dimensional example we could take the x_1 and x_2 axes, for example, as east-west and north-south axes intersecting at some convenient point. For a choice of (x_1, x_2) we would take a core through the layer of shale as a single drawing of material. We could draw either a simple random sample or a systematic sample using such a two-dimensional set of coordinates.

Once a sample has been drawn by randomization procedures from a population and the measurements of interest have been made on the objects in the sample, questions then arise as to how to estimate the population quantities or parameters of interest from the sample and how to determine the extent to which such estimates are subject to sampling fluctuations, that is, fluctuations that could be expected if repeated samples were drawn. Also, questions arise as to whether a sample could reasonably have come from a specified population or whether two samples could reasonably have come from the same or identical populations. It is beyond the scope of this paper to go into the details of statistical techniques for dealing with such questions. They can be found in many books on statistics. Examples of geological studies that illustrate the application of various statistical techniques for dealing with questions of this kind (as well as some statistical techniques discussed later in this paper) are one by Kahn (1956) on packing properties of sand-size sediments, and one by Nederlof (1959) on sedimentology.

Distributions of Measurements

Whether one is investigating the precision or accuracy of a measurement process or applying such a process to the different parts or pieces that make up a sample, or to the pool of all such parts or pieces, one has the job of making order and sense out of the resulting collection of measurements. As mentioned earlier, the basic device with which one begins in studying a collection of measurements from a random sample is the cumulative frequency distribution of such a collection, which, in either table or graph form, essentially shows the number or fraction or per cent of measurements less than or equal to each possible value that the measurement can have. It may be relevant here to make some comments about several particularly important types of frequency distributions that occur often in geology.

One of the most widely used measurements on samples from such geological populations as clay, sand, pebble deposits, rock bodies, etc., seems to be particle size and, of course, particle-size distributions for all particles in a sample unit or in a pool of sample units. It was observed a good many years ago by Loveland and Trivelli (1927) and more recently by investigators in various fields, including geology, that the logarithm of particle size in a sample of particles generated by grinding, crushing, and some other disintegrating actions has approximately a normal distribution, which, of course, can be characterized by two parameters, namely, the mean and standard deviation of logarithm of size. In geology, Krumbein (1934) first observed that the logarithm of particle sizes in clastic sediments have approximate normal distributions. Particle-size analyses have since been made by various geologists on many different geological materials with similar findings. References to many of these studies are given in a paper by Rogers (1960).

Kolmogorov (1941) has shown that the phenomenon of normality of logarithm of particle size is to be theoretically expected as a result of such actions, as continued grinding or crushing, on a collection of particles under certain conditions, including an assumption of homogeneous material, and that the mean and standard deviation for any particular distribution depend on physical characteristics of the material being ground or crushed and on the duration and other characteristics of the grinding and crushing process. In practice, these two parameters, that is, mean and standard deviation of the logarithm of particle size, have to be estimated experimentally from a sample of the particles produced by such a process.

Krumbein (1934) has devised a scale, called the ϕ-scale, for measuring particle sizes, which is now widely used in geology. If d is the diameter of a particle, in millimeters, the value of ϕ for the particle is defined by the relation $\phi = -\log_2 d$. Thus, increasingly small values of d correspond to increasingly large values of ϕ. The cumulative frequency distribution of values of ϕ for particles in a sample of particles is thus approximated by a cumulative normal distribution. This makes it quite easy to determine graphically the median, lower and upper quartile, and percentiles of grain sizes.

Cumulative particle-size distributions have been determined for many kinds of geological material, using sieving, hydraulic settling, thin-section, and other techniques for measuring sizes of particles. The curve obtained in each case is thus characteristic of the sample of particles. As one might expect, these curves vary widely from sample to sample not only from one kind of material to another but from one sample to another of the same material, depending on atmospheric, hydraulic, and other geological disintegrating processes that have acted on the material. If the sample consists of a mixture of particles produced by a mixture of two (or more) kinds of geological disintegrating actions, each of which would produce ϕ-values with normal distributions, the over-all particle-size distribution in ϕ units would be the sum of the two (or more) normal distributions and would, in general, be non-normal. There is a strong temptation (which occurs not only in geology but in biology and other fields) to make the converse statement, namely, that if a non-normal distribution exists it can be inferred that it is the sum of two or more normal distributions. This is a hazardous inference, since there could be homogeneous grinding, crushing, or sorting actions that do not produce normal particle-size ϕ distributions. To make any headway in verifying such a converse statement, one would need reasonably good information or strong grounds for some hypothesis as to how many different major grinding, crushing, or sorting actions were involved in producing the resulting sample of particles, and one would also need to know that each action would by itself produce a normal distribution of ϕ-values. One could then decompose the composite sample distribution into its normal components and estimate the fraction of particles involved in each component distribution. This, however, is a rather involved mathematical exercise, even for a mixture of two or three populations, which should be undertaken only when one has good information or a strong hypothesis that a small number of different distributions have been mixed. Such an analysis for a mixture of more than three or four distributions should be made, if at all, only with great caution.

It would appear that the most useful role a particle-size distribution can play is to characterize the results of a grinding or crushing or sorting action, or a combination of such actions, and to provide a basis for comparing the results of two or more such actions. Such distributions can provide basic information for classifying populations of materials from which the samples producing such distributions are drawn. This point will be discussed in the section "Classification and Discriminant Functions." Such a distribution can provide confirmatory evidence as to whether a mixture of two or more distributions of given kinds, each produced by a relatively homogeneous action, has occurred.

An interesting type of distribution analysis in geology on which I would like to make some comments is that used by Chayes (1956) for *modal analysis* in petrology. This consists of estimating from thin sections of the rock percentages of the major kinds of minerals that occur in rocks. Modal analysis has also been based on other procedures, such as counting crushed fragments of rocks and calculating areas covered by each species of mineral in polished slabs by planimetric methods. Chayes's method is basically a geometrical procedure for estimating the fraction of area in the thin section covered by each kind of mineral by counting, under a microscope, points in a lattice of points (covering a field of several hundred square millimeters) that fall into areas covered by each kind of mineral. The percentages thus found are estimates of percentage of volume occupied by each kind of mineral, which can be converted to percentages by weight by using differential weights proportional to specific gravities of the different kinds of minerals. Chayes has established what seems to be a quite satisfactory degree of precision of this method, particularly for the more abundant minerals, as compared, for example, with the degree of precision of petrochemcial determination of major elements in rocks.

The problem of determining the accuracy of Chayes's method of modal analysis is complicated by the fact that there seems to be no previously developed "standard" method with established accuracy for modal analysis. One approach to the accuracy of the method is to base it on the geometrical soundness of the procedure. A point that may be worth making on this approach is that, if the accuracy of the method is to be made to rest completely on geometric considerations, a careful study should be made of results that would be obtained if principles of randomization were strictly adhered to in both locating and orienting thin sections to be cut from rock samples. Such a study should be made on various kinds of rocks and various lattice spacings. Results obtained in this manner should be virtually free of any bias resulting from stratification, folding, foliation, and other geometric deviations from homogeneous mixing of mineral particles in the rock. If a rock consists of flat strata, the best direction for cutting sections, in terms of efficiency of unbiased estimation of percentages of major kinds of mineral, is perpendicular to the layers. However, if the strata are severely curved or folded, thus resulting in significant thickening or thinning of strata, attempts at cutting at right angles to stratification could result in some bias for purely geometric reasons.

Finally, it should be noted that perhaps the most common type of distribution arising in geology is the distribution of major chemical constituents in a rock expressed as weight percentages (or parts per million [p.p.m.] for the rare elements), such distributions being determined by chemical, spectrometric, photometric, X-ray, fluorescent, and other methods. As mentioned earlier, various studies have been made on the precision and accuracy of different methods of determining the chemical constituents of rocks, much of which seems to have been stimulated by the studies of Fairbairn *et al.* (1951) and of Stevens *et al.* (1960), conducted by the United States Geological Survey. Similar percentage distributions are obtained for minerals in a rock in modal analysis by thin-section and particle-sorting procedures. The percentage distribution of chemical constituents or minerals in a rock specimen is essentially a vector measurement whose components are non-negative and add up to 100, and it can be used as a device for numerically describing the rock. Table 1 gives such a distribution for G-1 based on 1951 data and for G-1 based on 1960 data, as well as similar distributions for W-1. Each of these four distributions is a vector with thirteen components. These percentage distributions are useful for rock classification purposes. Some comments on the use of such distributions in geological classification analysis will ge given in a later section.

Statistical Regression Methods

There is an increasing number of situations in geology in which linear statistical regression methods are being used. It may be useful to discuss these methods briefly, with particular reference to what is called "trend-surface" analysis in geology, in which the methods appear to be particularly promising.

First, let us discuss the simplest case of linear regression analysis, namely, that in which it is assumed that the value of a variable x for each object in a population is known or is readily obtainable, while values of a variable y of major interest may be more difficult to obtain, and where it is assumed that the two measurements x and y on each object in the population are such that the value of y for an object can be satisfactorily estimated from the value of x of that object as $\beta_0 + \beta_1 x$, where β_0 and β_1 are constants that must themselves be estimated from the values of (x, y) in a sample of objects from the population under study. Thus if $(x_1, y_1), \ldots, (x_n, y_n)$ are values of (x, y) in a sample of n objects, the errors $e_1, \ldots, e_n$ that would be made in using $\beta_0 + \beta_1 x_1, \ldots, \beta_0 + \beta_1 x_n$ as estimates for $y_1, \ldots, y_n$, respectively, in the sample in hand are $e_1 = (y_1 - \beta_0 - \beta_1 x_1), \ldots, e_n = (y_n - \beta_0 - \beta_1 x_n)$. As mentioned above, β_0 and β_1 are unknown and must be estimated from the (x, y) values in the sample. The usual procedure for estimating β_0 and β_1 is to choose the value of β_0 and of β_1 that make the sum of squared errors $e_1^2 + \ldots + e_n^2$ as small as possible. The details of this procedure, known as the *method of least squares*, are given in most statistics books and will not be given here. If we denote the estimates of β_0 and β_1 by b_0 and b_1 then $b_0 + b_1x$ becomes the estimator for the y-value of any object in the population if its x-value is known. The criterion by which it is judged whether the regression procedure outlined above has acceptable usefulness as a procedure for estimating the y-value of an object from its x-value is to compare the sum of squares

$$S_1 = (y_1 - b_0 - b_1x_1)^2 + \ldots + (y_n - b_0 - b_1x_n)^2$$

with the sum of squares

$$S_2 = (y_1 - \bar{y})^2 + \ldots + (y_n - \bar{y})^2 ,$$

where $\bar{y}$ is the average of $y_1, \ldots, y_n$. If x provides no information at all for linearly estimating y, then S_1 will not be significantly smaller than S_2 and the points $(x_1, y_1), \ldots, (x_n, y_n)$ will be randomly scattered about the horizontal line having equation $y = \bar{y}$ in the xy-plane. If x provides perfect information for linearly estimating y, then S_1 will be zero and all the points $(x_1, y_1), \ldots, (x_n, y_n)$ will lie on the straight line having the equation $y = b_0 + b_1x$ in the xy-plane. In any intermediate case the sample points $(x_1, y_1), \ldots, (x_n, y_n)$ will be scattered about the *regression line* of y on x having the equation $y = b_0 + b_1x$. If the scatter of points is sufficiently close to the regression line to consider $b_0 + b_1x$ a useful estimator for y for all objects in the sample, then it is assumed that, since the sample is drawn at random from the population by randomization procedures, $b_0 + b_1x$ will also provide useful estimates of y for the objects in the population that are not contained in the sample.

There are many examples in geology of applications of this simplest case of linear regression. For instance, Finkel (1959) has made a study of barchans (a crescent-shaped sand dune) in southern Peru in which he has devised linear regression schemes for estimating width (y) across barchan horns as a function of height (x) of slip face of barchans, as well as other characteristics of the barchans as functions of height of their slip face. Sarmiento (1961) has made a regression study for linearly estimating porosity (y) of rocks from acoustic velocity (x) of the rocks. Weber and Middleton (1961) have made regression studies for linearly estimating per cent cobalt (y) from per cent nickel (x) in turbidites. Taylor, Sachs, and Cherry (1961) have made regression analyses for estimating per cent alumina (y) from per cent silica (x) and per cent magnesia (y) from silica (x) in tektites.

The concepts underlying in the simple linear regression model involving only one x variable extend in a straightforward manner to the more general case in which there are several readily available x variables (i.e., a vector), say $x_1, \ldots, x_k$, on each object in the population. Here it is desired to devise a scheme for linearly estimating the value of y of an object from values of $x_1, \ldots, x_k$ of the object. More precisely, suppose we have a sample of n objects from the given population drawn by randomization procedures. Let the values of $(x_1, \ldots, x_k, y)$ for the n objects in the sample be $(x_{11}, \ldots, x_{k1}, y_1), \ldots, (x_{1n}, \ldots, x_{kn}, y_n)$ and assume that the y of an object can be satisfactorily estimated from its $x_1, \ldots, x_k$ as $\beta_0 + \beta_1 x_1 + \ldots + \beta_k x_k$ for suitably chosen values of $\beta_0, \beta_1, \ldots, \beta_k$. Thus, for the n objects in the sample, the errors $e_1, \ldots, e_n$ that would be made if we used as estimates of $y_1, \ldots, y_n$ the quantities

$$\beta_0 + \beta_1 x_{11} + \ldots + \beta_k x_{k1}, \ \ldots, \ \beta_0 + \beta_1 x_{1n} + \ldots + \beta_k x_{kn},$$

respectively, would be given by

$$e_1 = (y_1 - \beta_0 - \beta_1 x_{11} - \ldots - \beta_k x_{1n}), \ \ldots, \quad e_n = (y_n - \beta_0 - \beta_1 x_{1n} - \ldots - \beta_k x_{kn}),$$

respectively. As in the case of simple linear regression involving only one x, the "best" estimates of $\beta_{,0}$ $\beta_1, \ldots, \beta_k$, say $b_0, b_1, \ldots, b_k$, are the values of $\beta_0, \beta_1, \ldots, \beta_k$ that make the sum of squared errors $e_1^2 + \ldots + e_n^2$ as small as possible. The procedure for minimizing this sum of squares, and for finding the values of $b_0, b_1, \ldots, b_k$ in terms of values of $(x_1, \ldots, x_k, y)$ on the n objects in the sample, can be found in most statistics books and will not be given here. The problem of computing the values of $b_0, b_1, \ldots, b_k$ from the sample is completely routine; in fact, programs exist for doing the computing on various electronic computers. The essential point is that from the sample one obtains a device for linearly estimating the value of y of an object in the sample from the values of $x_1, \ldots, x_k$ of that object, the estimator for y being $b_0 + b_1 x_1 + \ldots + b_k x_k$. As in the simple regression case, the basic criterion for judging the effectiveness of the estimation is to compare the sum of squares

$$S_1 = (y_1 - b_0 - b_1 x_{11} - \ldots - b_k x_{1n})^2 + \ldots + (y_n - b_0 - b_1 x_{1n} - \ldots - b_k x_{kn})^2$$

with the sum of squares

$$S_2 = (y_1 - \bar{y})^2 + \ldots + (y_n - \bar{y})^2,$$

both of which can be calculated from the sample. If $x_1, \ldots, x_k$ contain no information for estimating y, S_1 will not be significantly smaller (statistically) than S_2; if they contain perfect information for linear estimation of y, S_1 will be zero; if they contain enough information for satisfactory linear estimation of y, S_1 will be "appropriately" small compared with S_2. If, upon comparing S_1 with S_2, it is decided that

$b_0 + b_1x_1 + \ldots + b_kx_k$ is a useful estimator for y for the objects in the sample, it is assumed that, since the sample has been drawn from the population by appropriate randomization procedures, the validity of the estimator extends to objects in the population that are not contained in the sample. In some fields of inquiry by regression analysis it is customary to run a check on the validity of $b_0 + b_1x_1 + \ldots + b_kx_k$ as an estimator for y by drawing a second sample of objects from the population and checking how well $b_0 + b_1x_1 + \ldots + b_kx_k$ estimates y of each object in the second sample, where $b_0, b_1, \ldots, b_k$ have been determined from the first sample. One can also interchange the roles of the two samples and estimate y values of the first sample from information contained in $x_1, \ldots, x_k$ of the second sample. Such a procedure is sometimes called a *two-sample validity cross-check* of the regression procedure.

It should be pointed out that there are statistical techniques available in various statistics books for deciding whether or not the inclusion of additional x variables, say $x_{k+1}, \ldots, x_r$, will significantly improve an estimator for y that has been based only on the variables $x_1, \ldots, x_k$.

Various regression studies in geology involving two or more x variables have been made. For instance, Kahn (1956), in his study of packing properties of grain-size sediments, has used two variables, namely, packing density (x_1) and intercept size (intersection of grained traverse) (x_2) for estimating packing proximity (y) of sediments. Krumbein (1959) has made a regression analysis of beach firmness in which he linearly estimates beach firmness (y) from mean grain size (x_1), degree of sorting of grains (x_2), moisture content (x_3), and porosity (x_4).

Some of the most interesting and promising applications of the more general linear regression model in geology are in the fitting of *trend surfaces* for various geological variables over a geographical area. For instance, suppose one is interested in the distribution of gravity, per cent iron, per cent alumina, mean particle size, or any other geological variable over a given region, which may be of an order ranging from hundreds of square yards to hundreds of square miles, depending on the problem under study. If we set up a plane over the region with rectangular axes, which we may call the uv-axes, and if at any point (u, v) in the region we let y be the value of the geological variable of interest, we may think of the value of y as an ordinate erected perpendicular to the uv plane. We would have some kind of mathematical surface over the region giving y as a function of (u, v). This surface will be called the *trend surface* for y. The objective in trend-surface analysis is to find a practically useful *approximation* to this surface from information consisting of measurements on y at each of a sample of n suitably chosen points in the uv-plane. Such an approximating surface could then be used for estimating the value of y at points in the uv-plane that are not contained in the sample. For convenience, we may take the n points as intersection points of a rectangular or square grid. (It is not necessary, however, to have the n points arranged in the form of a grid. They can be distributed over the area in an irregularly spaced manner, although the computation will, in general, be simplified if the points are arranged in a rectangular grid.) Let $(u_1, v_1), \ldots, (u_n, v_n)$ be the coordinates of the n points thus selected. The spacing of the sample grid points depends on the problem at hand, the main principle being that the fluctuations in the variable y from one grid point to an adjacent one should, in general, be relatively small compared with fluctuations of y over the whole region being sampled. If there is uncertainty about the validity of the assumption that fluctuations in y between adjacent sampling points are small relative to those over the region being sampled on account of insufficient prior geological knowledge about variability of y, a preliminary study may be worthwhile to examine the variability of y before undertaking the trend-surface fitting study. A conservative approach, with only a small amount of information on the variability of y available, would be to tend to select a spacing that might be too small rather than too large.

At each of these n points thus chosen the value of y is determined, the n values of y being denoted by $y_1, \ldots, y_n$, respectively. It is now assumed that the trend surface can be satisfactorily approximated by a polynomial whose equation is of form

$$y = \alpha_{00} + \alpha_{10}u + \alpha_{01}v + \alpha_{20}u^2 + \alpha_{02}v^2 + \alpha_{11}uv + \ldots$$

up to terms in u and v of second or third or possibly higher degree. For simplicity, suppose we consider the case of terms as far as second degree. If we denote u, v, u^2, v^2, uv by x_1, x_2, x_3, x_4, x_5, respectively, and α_{00}, α_{10}, α_{01}, α_{20}, α_{11} by β_0, β_1, β_2, β_3, β_4, β_5, respectively, then the equation of the approximating trend surface is $y = \beta_0 + \beta_1 x_1 + \ldots + \beta_5 x_5$, where $\beta_0, \beta_1, \ldots, \beta_5$ have to be determined from the sample of n points. Thus we have a typical linear regression problem. For the jth object in the sample we have the observed values (u_j, v_j, y_j), or, expressed in terms of x's, we have $(x_{1j}, \ldots, x_{5j}, y_j)$, where, of course,

$$x_{1j} = u_j,\ x_{2j} = v_j,\ x_{3j} = u_j^2,\ x_{4j} = v_j^2,\ x_{5j} = u_j v_j\ .$$

Estimates $b_0, b_1, \ldots, b_5$ of $\beta_0, \beta_1, \ldots, \beta_5$ are those values of $\beta_0, \beta_1, \ldots, \beta_5$ for which the sum of squared errors

$$(y_1 - \beta_0 - \beta_1 x_{11} - \ldots - \beta_{51} x_{51})^2 + \ldots + (y_n - \beta_0 - \beta_1 x_{1n} - \ldots - \beta_5 x_{5n})^2$$

has its smallest value. Hence the approximating trend surface would have as its equation

$$y = a_{00} + a_{10}u + a_{01}v + a_{20}u^2 + a_{02}v^2 + a_{11}uv\ ,$$

where $a_{00} = b_0$, $a_{10} = b_1$, $a_{01} = b_2$, $a_{20} = b_3$, $a_{02} = b_4$, $a_{11} = b_5$. The problem of judging the effectiveness of this approximating trend surface is very tricky and laden with complications. We first assume that evidence exists that shows that the spacing of the sample points in the grid has been of such a magnitude that fluctuations of values of y between adjacent sample points are relatively small compared with fluctuations of y over the region of interest. Under this assumption the question of how good the approximating trend surface is reduced to a matter of comparing the sum of squares of differences (residuals) $S_1 = (y_1 - y_1')^2 + \ldots + (y_n - y_n')^2$, where y_j' is the estimate of y_j given by

$$y_j' = a_{00} + a_{10}u_j + a_{01}v_j + a_{20}u_j^2 + a_{02}v_j^2 + a_{11}u_jv_j\ ,$$

with the sum of squares $S_2 = (y_1 - \bar{y})^2 + \ldots + (y_n - \bar{y})^2$, where $\bar{y}$ is the mean of $y_1, \ldots, y_n$, the sample values of y. The smaller the value of S_1 relative to S_2, the more closely the approximating trend surface fits the n sample points $(u_1, v_1, y_1), \ldots, (u_n, v_n, y_n)$. On the other hand, if we take more and more terms involving higher powers of u and v in the equation of the approximating trend surface, we rapidly increase the number of constants $a_{00}, a_{10}, a_{01}, \ldots$ to be determined from the data and consequently obtain a better and better fit to the n sample points. Thus, if a polynomial in u and v of degree k is fitted, the number of constants to be determined is $\frac{1}{2}(k+1)(k+2)$. In the extreme case, if we take n terms in the equation to be fitted, we would get a perfect fit to the sample points. This does not necessarily mean that we would get a good fit to values of y at points in the uv-plane interspersed among the n sample points. If the number of terms in the equation of the surface to be fitted is quite large relative to the number of sample points—or even if the number is not so large, for that matter—it may be useful to make a two-sample validity cross-check by taking a second grid sample of points and finding out how well the approximating trend surface determined by the first sample fits the y-values of the second sample and vice versa. More specifically, the grid for the second sample may be obtained by moving the grid of the first sample one-half of a spacing parallel to the u-axis and one-half of a spacing parallel to the v-axis. Thus, if we call the two samples A and B, there are statistical techniques for comparing and interpreting the sums of squares of the four sets of "errors" or residuals between y-values and estimated y-values one obtains from the two samples; two sets being those produced by applying estimators computed from sample A to estimate y-values in sample A, and those obtained by applying estimators computed from sample A to estimate y-values from sample B; the other two sets being the corresponding sets obtained by interchanging A and B. The essential condition to be satisfied by these four sets of residuals, stated roughly, is that the sums of squares for each of the four sets must be relatively small compared with the sum of squares $S_1 = (y_1 - \bar{y})^2 + \ldots + (y_n - \bar{y})^2$ for either sample. If the sum of squared residuals obtained when applying estimators for y computed from sample A to sample B (or vice versa) are not small relative to S_1, one runs the risk of having a "spuriously close" approximating trend surface, that is, one that will fit well at the sample points but will not provide a generally good fit over the region. After an approximating trend surface has been fitted, one should examine the sign and magnitude of the residuals $(y - y')$ over the sample points, where y is the observed value of y and y' the estimated value of y. Patches of positive and/or patches of negative residuals can usually be taken as evidence that not enough terms have been taken in the equations of the approximating trend surface.

Once a satisfactory approximating trend surface has been fitted—satisfactory in the sense of meeting conditions such as those described above—it can be used for constructing contour lines of constant value of y to give a graphical picture of how y varies over the region being studied.

One of the earliest studies of trend-surface fitting in geology seems to be one by Simpson (1954), who fitted polynomials up to degree four in two variables to gravity data. Grant (1957) has also applied the technique to gravity data obtained over known sulphide deposits in Quebec. Miller (1956) has applied the method to determine an approximating trend surface to terms of second degree of median sediment size (and also to standard deviation of sediment size) over a near-shore region just off La Jolla, California, using sediment parameters (medians and standard deviations in ϕ-units) of batches of sediment analyzed by Inman (1953) from each of some 140 sampling points in the region. Krumbein (1959) has applied trend-surface approximation analysis, using polynomials up to the third degree, to the clastic ratio of Pennsylvanian rocks in a large area covering parts of Kansas and Oklahoma. Whitten (1961) has used the method for fitting approximating trend surfaces of color index, per cent quartz, per cent hornblende, per cent sphene, and per cent of several other constituents of a granitic mass in County Donegal, Eire. Some investigators who have used these methods have not, in the author's opinion, been sufficiently rigorous in testing whether their fitted surfaces satisfactorily meet the criteria discussed in the preceding paragraph or other similar criteria.

There is one more point about trend-surface analysis that may be worth some attention. Suppose the objective in a trend-surface analysis is to locate a small area within the geographical region under study in which the variable y is a maximum (or a minimum, for that matter). Such a situation could arise, for instance, in searching for the location of high concentrations of a mineral distributed over a region. If the variable y can be readily measured in the field after a sample has been taken at any point (u, v), then there are statistical procedures for selecting a sequence of sampling points in a grid so as to reach the area of maximum (or minimum) concentration of y using a minimum number of sampling points in the grid. The procedure essentially amounts to fitting relatively simple approximating trend surfaces over small areas sequentially, so that after each fitting one determines which grid points to take for the next fitting step. This makes it unnecessary to try to fit approximating trend surfaces over the entire region. If measuring y at a point (u, v) requires laboratory procedures that cannot be carried out at the site, then one might gather samples of material over an entire grid of sampling points in the uv-plane covering the region of interest, with the view of taking them back to the laboratory for analysis. One would actually measure y sequentially over certain sample points in the grid so as to reach the maximum (or minimum) of y with a minimum number of measurements on y. Such procedures have been devised by Box (1954) for use in chemical and other continuous production process industries for finding combinations of two or more controllable variables that maximize yield or some other variable of major interest in mass production. Geologists interested in the problem of locating within a given region small areas in which y has a maximum (or a minimum) might find it worthwhile to examine the possibilities of this sequential method.

Analysis of Variance

One of the more sophisticated techniques of statistical analysis and inference that one finds appearing with increasing frequency in geological studies is that known as *analysis of variance.* This technique, used in conjunction with carefully designed experiments, can be very useful in studying the effects of various factors on the variability of measurements of geological variables.

The basic idea of analysis of variance is the recognition of the fact that the variability of any important variable y from object to object in a population usually depends on factors that can be used for classifying the objects of the population into subpopulations, within each of which the variability of y is or may be significantly reduced. Conceptually, the main objective of analysis of variance is to break up the variance of y over the entire population (that is, to break up the average of $(y - \mu)^2$ over all objects in the population where μ is the mean of y's of all objects in the population) into components associated with the major factors contributing to the variance of y. As usual, it is not possible or feasible to measure y for all objects in the population, and hence a sample must be designed and its elements drawn from the various subpopulations into which the population has been subdivided so that the relevant components of the population variance can be estimated from the information contained in the sample. In many problems it is not known a priori whether the factors under study contribute significantly to the variability of y, in which case a statistical test known as the *variance ratio test* can be made on the basis of information in the sample as to whether such factors do, in fact, contribute significantly. The problem of designing such samples so that they will efficiently provide information about the components of variance with a minimum of computational complications is a specialized and sophisticated branch of statistics commonly referred to as *design of experiments*. The principles of design of experiments and the application of analysis of variance techniques to measurements obtained from such experiments can be found in a number of statistics books, including those by Bennett and Franklin (1954), Cochran and Cox (1957), Cox (1958), Fisher (1935), and, specifically, for problems in the geological sciences, in the book by Miller and Kahn (1962). No attempt will be made to go into the details of the methods here. It will perhaps be sufficient to mention several examples in geology in which these methods have been used and to make some comments about them.

A study of the homogeneity of size distributions of wave-deposited particles of several different minerals on a beach has been made by McIntyre (1959), using analysis of variance techniques. In his study McIntyre placed a rectangular grid of five rows and ten columns, ruled into 1 × 1-foot squares, onto the Lorraine Beach on Lake Erie so that the first row was along the shore line, while the fifth row was five feet from the shore line. At each intersection on the beach a batch of 2 ml. of material was taken from each of three well-defined layers of sand called A, B, and C (top, middle, and bottom). By successively splitting each of the resulting 150 batches of sand with an Otto microsplit, enough sand was taken so that when spread upon a miscroscope slide it could be covered by a 30 × 20-mm. cover glass. By using a randomization procedure for selecting grains, 10 grains of quartz, and 5 grains each of garnet, hornblende, clino-pyroxene, and hypersthene were selected. The long axis of each of the 4,500 thus selected was measured. Thus, as far as quartz is concerned, McIntyre's experiment is a tenfold replication of a complete 3 × 5 × 10 factorial experiment and a fivefold replication of such an experiment for each of the other four minerals. The general design of the 3 × 5 × 10 factorial experiment is shown in table 2.

For each type of grain an analysis of variance was made to determine the effect due to row, due to column, and due to row by column interaction, with variation of lengths of grains of each mineral within the slides as the yardstick ("error") to measure the various effects for each mineral. The net result of this study was that all except hypersthene grains were homogeneously deposited row by row (perpendicular to shore line) in the A layer, and all grains except hornblende were homogeneously deposited in both directions in the B layer. In the C layer, however, deposition was significantly non-homogeneous column by column (along the beach) in all minerals except hypersthene, while deposition was homogeneous row by row (perpendicular to shore line) except

TABLE 2

Design of Complete $3 \times 5 \times 10$ Factorial Experiment Used by McIntyre in His Study of Size Distribution of Wave-deposited Particles of Quartz, Garnet, Hornblende, Clino-pyroxene and Hypersthene at Lorraine Beach on Lake Erie

Rows (Levels Perpendicular to Beach)	Layer	Columns (Levels Parallel to Beach)									
		1	2	3	4	5	6	7	8	9	10
1	A B C										
2	A B C										
3	A B C										
4	A B C										
5	A B C										

for hornblende. The author then proposes interpretations of these effects in terms of hydraulic action on the grains.

An example of a more complex experimental design and its corresponding analysis of variance has been carried out by Flanagan (1960). In his study he investigated the effect of various factors on the variability of lead-content determinations (in p.p.m.) in the well-known collaborative petrochemical study of granite G-1 conducted by the United States Geological Survey. Six bottles of G-1 were halved, rebottled, and numbered 1, 2, . . . , 12. Bottles 1 and 7 formed batch I, bottles 2 and

8 batch II, and so on, thus giving batches I, II, III, IV, V, and VI. In the first round, laboratories A, B, C, D, E, and F made lead content determinations from each pair of bottles in batches I, II, III, IV, V, and VI, respectively. Then the batches were rotated so that, in the second round, laboratories B, C, D, E, F, and A made lead content determinations from each pair of bottles in I, II, III, IV, V, and VI, respectively. And so on. After six rounds, seventy-two determinations of lead content had been made, so the results can be exhibited as a duplicated 6 × 6 Latin Square experimental design, as shown in table 3.

TABLE 3

DUPLICATED 6 × 6 LATIN SQUARE EXPERIMENTAL DESIGN TO STUDY SOURCES OF VARIATION IN LEAD CONTENT DETERMINATIONS FROM GRANITE G-1 (DATA FROM FLANAGAN, 1960)

Batch No.	Bottle No.	Round No.					
		1	2	3	4	5	6
I	1	A57	B50	C49	D46	E49	F47
	7	60	43	47	51	43	48
II	2	B50	C49	D46	E47	F47	A55
	8	48	48	56	48	48	60
III	3	C47	D44	E46	F48	A53	B47
	9	47	46	47	48	55	53
IV	4	D51	E47	F47	A60	B42	C49
	10	46	50	47	54	45	47
V	5	E49	F47	A56	B53	C46	D48
	11	46	45	56	58	38	50
VI	6	F46	A56	B52	C48	D44	E49
	12	47	46	50	43	46	46

The analysis of variance for the lead content measurements in table 3 is routine, and for the details of the analysis the reader is referred to Flanagan's paper. The essential point to be noted here is that the experiment was designed with considerable sophistication so that the effects of four different factors on the variability of lead content measurements could be isolated and estimated, namely, effect due to laboratories, effect due to batches, effect due to order of analysis (rounds), and effect due to duplication (bottles), with the effect due to duplication being used as the yardstick ("error") for measuring the effect due to each of the other three factors. The analysis of variance established that the factor of laboratories had, by far, the greatest effect on variability in the measurements, while the effect of order of analysis also had a significantly large effect. The magnitude of the effect due to batches was approximately the same as that due to duplication ("error"). The problem of breaking the large laboratory-to-laboratory variation into subeffects due to analysts, variations in laboratory procedure, etc., was not undertaken, but any effort to reduce variability of lead content determinations in such interlaboratory studies as this one must be focused on the tightening of standards from laboratory to laboratory. Incidentally, one of the outcomes of this study was the establishment of 50 p.p.m. as lead content of G-1 rather than the value of 27 p.p.m., which had been recommended on the basis of earlier determinations in 1954.

Further examples of applications of analysis of variance techniques to geology problems are the following: Kahn (1956) studied the variation of packing properties of particles in three different rock types (arkose, graywacke, and quartzite), each from three different formations, by using two slides from each formation and running five traverses on each slide. Krumbein and Tukey (1956) studied the spatial variability of pebble counts in various drifts, of percentages of heavy minerals in beach sands of Lake Michigan, and of other quantities. Griffiths (1959) has made a study of the sources of variability of long-axis measurements of rock fragments from Fishing Creek, near Lamar, Pennsylvania.

Classification and Discriminant Functions

Linear regression and analysis of variance techniques are examples of multivariate statistical techniques. These methods, as we have seen, have been used to a limited extent in geological studies. There is a further multivariate statistical method, commonly called *discriminant analysis*, which should be useful in some problems of classification in geological studies. Basically, the method is a procedure for separating the objects in a mixture of two or more closely related varieties back into the component varieties on the basis of a vector of measurements made on each object in the mixture. It has been used on such problems as classifying skulls as belonging to one of two or more races, selection of desirable strains in animal or plant breeding, assigning authorship of the *Federalist* papers to Hamilton or Madison, etc. Very little use seems to have been made of discriminant analysis in the geological sciences yet. Miller and Kahn (1962) mention several examples. However, I have noted many instances in which the basic idea has been used informally in simple form. For example, Kagami (1961) has shown that by taking two parameters of a particle-size distribution—namely, x_1 = median particle size, and x_2 = skewness—submarine sediments taken off Sakata, Japan, can be quite well classified by values of x_1 and x_2. More precisely, he finds values of (x_1, x_2) for several batches of submarine sediment from each of six different types of sea bottom—upper shelf, mid-shelf, shelf margin, delta, near shore, and slope insular shelf—and shows that when these values of (x_1, x_2) are plotted in an x_1x_2 plane they fall in regions that are virtually non-overlapping. Thus, each of these six different types of sea bottom in the area studied is represented by a different region in the x_1x_2 plane.

There is, of course, an abundance of examples in which rocks are classified by their petrochemical percentage distributions, such as those discussed earlier and exemplified in table 1, or by their modes (percentage distributions of mineral constituents), as discussed, for instance, by Chayes (1956). It is difficult to represent such distributions graphically if there are more than three or four constituents. In the case of three constituents, I, II, and III, say, let the percentages of I, II, and III be x_1, x_2, and x_3, respectively, where $x_1 + x_2 + x_3 = 100$. An example familiar to geologists is that in which I, II, and III refer to sand, silt, and clay in particle-size analysis of sediment and where x_1, x_2, and x_3 are percentages of particles in the three classes. The three numbers (x_1, x_3, x_2) for any specific percentage distribution can be represented as a point in an equilateral triangle whose vertices are labeled I, II, and III, respectively. The scale for x_1 runs from 100 at I to 0 along the line II–III. Scales for x_2 and x_3 are similarly established. If we have n such distributions, we will have a cluster of n points in the triangle. Thus, if we have two types of rocks, say A and B, and if we have n_1 petrochemical percentage distributions from rock specimens of type A and n_2 such percentage distributions from rock specimens of type B, such that the nj points corresponding to the distributions from type-A rock fall into one region of the triangle, while the n_2 points corresponding to the distributions from type-B rock fall into another region in the triangle that overlaps little if any of the first region, we have a simple device for geometrically classifying rock types A and B. Note that one or all of the constituents I, II, and III may be obtained by combining constituents in a percentage distribution with a larger number of constituents or by expressing the percentages of three major constituents that do not add to 100 as percentages of their total.

As all geologists know, there are many examples in geological literature in which such classification triangles are used. For some recent examples of such classification triangles, the reader is referred to Ōki (1961), Stehli and Hower (1961), Weber and Middleton (1961), and Warshaw and Roy (1961).

The idea can be extended to the case of four constituents I, II, III, and IV with percentages x_1, x_2, x_3, x_4, where $x_1 + x_2 + x_3 + x_4 = 100$. Any specific distribution of these four constituents can be represented as a point in a regular tetrahedron whose vertices are labeled I, II, III, and IV and where $x_1 = 100$ at I and $x_1 = 0$ on the opposite face II–III–IV, with scales for x_2, x_3, x_4 being similarly defined. Examples of such classification tetrahedra are quite common in geological literature. Thus, if distributions on several analyses of each of two rock types, A and B, should plot into points in different regions in the tetrahedron, we would have a relatively simple scheme for geometrically classifying the two types of rocks.

Examples of such classification tetrahedra are familiar to geologists, but recent examples may be found in papers by Francis (1958), Mumpton and Roy (1958), and Ōki (1961).

The problem of extending the notion of a classification triangle or tetrahedron to the case of a percentage distribution having more than four constituents, if and when there is a need to do so, is geometrically awkward, since the geometry of a simplex (generalized tetrahedron) becomes rather complicated in more than three dimensions. But, algebraically, the problem can be handled by a method known as the *discriminant analysis*. This method has wider applicability than simply to petrochemical and other percentage distributions. I shall describe the main ideas of this method without going into technical details.

To describe more generally the kind of situation in which discriminant analysis may be useful, suppose we have a population of objects consisting of a mixture of two subpopulations, A and B, such that only by difficult and time-consuming analysis of some kind can one determine whether any given object from the over-all population belongs to A or B. But suppose one can find several measurements, say $x_1, \ldots, x_k$, which can be made quickly and objectively on each object. In the case of a petrochemical per cent distribution, $x_1, \ldots, x_k$ would be percentages. Suppose further that a sample of n objects is taken at random from the over-all population and that the values of $x_1, \ldots, x_k$ are determined for each of these objects. Let $(x_{11}, \ldots, x_{k1}), \ldots, (x_{1n}, \ldots, x_{kn})$ be the values of $(x_1, \ldots, x_k)$ on the n objects in the sample. The n vectors $(x_{11}, \ldots, x_k), \ldots, (x_{1n}, \ldots, x_{kn})$ can be thought of as n points plotted in a k-dimensional space with $x_1, \ldots, x_k$ as its axes. Suppose the n objects have been carefully analyzed so that it is known which of the n points correspond to A-type objects and which correspond to B-type objects. Now suppose one can find a direction in which to draw a straight line in this k-dimensional space such that, if the n points are projected onto the line perpendicularly, the projections of the n points on the line fall into two separate clusters, one of which consists of points projected from A-type points in the k-dimensional space and the other consists of points projected from B-type points. Then the A- and B-type objects have been separated along an ordinary one-dimensional line. Algebraically, this means determining quantities $a_1, \ldots, a_k$ from the sample $(x_{11}, \ldots, x_{k1}), \ldots, (x_{1n}, \ldots, x_{kn})$ so that the linear quantities $(a_1x_{11} + \ldots + a_kx_{k1}), \ldots, (a_1x_{1n} + \ldots + a_kx_{kn})$ would have values $z_1, \ldots, z_n$, respectively, which, if plotted on a z-axis, would fall into two clusters of points, one belonging to A-type objects and the other to B-type objects. There is a straightforward statistical procedure for finding values of $a_1, \ldots, a_n$ that will produce the "best" separation of values of the z's for A-type and B-type objects. One can then see on what part of the z-line points corresponding to A-type objects fall and on what part points corresponding to B-type objects fall. The inference is that if any object (not in the sample) is taken from the population and if $(x_1, \ldots, x_k)$ is measured on that object, it can be plotted on the z-axis (discriminant line) and declared an A-type object or a B-type object, depending on where it falls on the z-axis.

The simplest case of discriminant analysis occurs for $k = 2$, in which case the n points of the sample can be plotted as n points in the x_1x_2-plane. If each point is indicated as an A or B by different-colored points, for example, one can see whether the n points fall into two non-overlapping clusters—an A cluster and a B cluster. Even though there may be failure to separate into two such clusters in two dimensions, it is possible that separation could be produced by considering an additional variable x_3 (or several additional variables), thus going into three-or-more-dimensional discriminant analysis.

Discriminant analysis such as that described would be used only if discrimination cannot be achieved by using one (or two or a smaller number) of measurements. In practice, it is not always easy to find measurements $x_1, \ldots, x_k$ by which one can make a clean-cut discrimination between A-type and B-type objects by discriminant analysis—there is usually some overlap of A-type points and B-type points. But frequently one can do a reasonably good job of separation by such methods, using as few as two measurements x_1 and x_2. Thus, as mentioned earlier, Kagami (1961) could achieve almost perfect classification of six types of sea bottom off Sakata, Japan, by using only two measurements (x_1, x_2) on particle-size distributions of sediments taken from the several types of sea bottom, where x_1 is median particle size and x_2 is skewness.

It is likely that discriminant analysis in its simplest form, involving two variables, or in more sophisticated form, using three or more variables, will be found useful not only in such classification problems as those considered by Kagami but also in such problems as discriminating between two similar and coexisting species of fossils, or discriminating between similar igneous rocks from petrochemical or modal analyses, or discriminating between two similar types of rocks or material by using two or more percentiles in particle-size distributions.

The method of discriminant analysis can be extended to the problem of discriminating between three or more similar types of objects mixed up in an over-all population. The reader interested in the technical details of discriminant analysis and its applications to classification problems is referred to Rao (1952), which also contains a list of references on this method. The technical details of applying the method are given in the book by Miller and Kahn (1962).

Summary and Conclusions

During the last ten or fifteen years there has been a fairly strong growth in the application of statistical techniques to problems in geological research. Judging from this growth, these methods are apparently headed for considerably wider application in the future. Statistical techniques are being applied to such problems as the refinement of measurement processes, particle-size analysis, the study of relationships between geological variables, and the estimation of some variables from known values of others; to the study and isolation of portions of variability of geological variables attributable to various identifiable factors; and to quantitative classification of sediments, rocks, and other geological entities.

In this paper I have attempted to discuss principles and the main ideas of the statistical methods being used in these studies, without attempting to go into technical details. First to be discussed were the basic concepts underlying all statistical analysis and inference, such as precision and accuracy of measurement processes, statistical populations, sampling, randomization, and distributions in a context of geology problems. Then certain statistical techniques that appear particularly promising in geology were concentrated on. Thus, the ideas of linear regression analysis, with special reference to its application in the fitting of trend surfaces to geological variables over geographical regions, were presented. Next to be described was the analysis of variance as a technique for decomposing the variability of geological variables into components associated with specified sources of variability; the technique is particularly effective for analyzing the results of carefully designed factorial experiments. Examples of application of analysis of variance methods in geology were presented. Finally, a discussion was given of concepts of quantitative classification of geological entities by vectors of measurements and of discriminant analysis as a statistical quantitative classification device. Throughout the paper I have tried to stress the more important assumptions and limitations of the techniques, again without going into technical details. References to books giving the details are given at appropriate places throughout the paper.

No attempt has been made to discuss techniques for analyzing orientation measurements, time series, and other more highly specialized methods particularly applicable to geophysical problems.

Acknowledgments.—The author wishes to take this opportunity to express his appreciation to Dr. F. Chayes of the Carnegie Institution of Washington, Dr. F. J. Flanagan of the United States Geological Survey, Dr. J. S. Kahn of the Lawrence Radiation Laboratory of the University of California, Professor W. C. Krumbein of Northwestern University, and Professor D. B. McIntyre of Pomona College for suggestions that improved the manuscript. Any inaccuracies, errors, and other defects that remain are mine, and not theirs.

References Cited

Ahrens, L. H., and Taylor, S. R., 1961, Spectrochemical analysis, 2d ed.: Reading, Mass., Addison-Wesley Pub. Co.

Baird, A. K., MacColl, R. S., and McIntyre, D. B., 1961, A test of the precision and sources of error in quantitative analysis of light, major elements in granitic rocks by X-ray spectrography: Advances in X-ray analysis, v. 5, p. 412–422.

Bennett, C. A., and Franklin, N. L., 1954, Statistical analysis in chemistry and the chemical industry: New York, John Wiley & Sons.

Box, G. E. P., 1954, The exploration and exploitation of response surfaces: some general considerations and examples: Biometrics, v. 10, p. 16–60.

Chayes, F., 1956, Petrographic modal analysis: New York, John Wiley & Sons.

Cochran, W. G., 1953, Sampling techniques: New York, John Wiley & Sons.

——— and Cox, G. M., 1957, Experimental designs, 2d ed.: New York, John Wiley & Sons.

Cox, D. R., 1958, The planning of experiments: New York, John Wiley & Sons.

Dennen, W. H., Ahrens, L. H., and Fairbairn, H. W., 1951, Spectrochemical analysis of major constituent elements in rocks and minerals, pt. 3 *of* A cooperative investigation of precision and accuracy in chemical, spectrochemical, and modal analysis of silicate rocks: U.S. Geol. Survey Bull. 980. p. 25–45.

Fairbairn, H. W., *et al.*, 1951, A cooperative investigation of precision and accuracy of chemical, spectrochemical, and modal analysis of silicate rocks: U.S. Geol. Survey Bull. 980.

Finkel, H. J., 1959, The barchans of southern Peru: Jour. Geology, v. 67, p. 614–647.

Fisher, B. A., 1935. The design of experiments: Edinburgh, Oliver & Boyd.

Flanagan, F. J., 1960, The lead content of G-1, pt. 5 *of* Second report on a cooperative investigation of the composition of two silicate rocks: U.S. Geol. Survey Bull. 1113, p. 113–121.

Francis, G. H. M., 1958, Petrological studies in Glen Urquhart, Inverness-shire: Bull. British Museum, v. 1, p. 123–162.

Grant, F., 1957, A problem in the analysis of geophysical data: Geophysics, v. 22, p. 309–344.

Griffiths, J. C., 1959, Size and shape of rock fragments in Tuscarora scree, Fishing Creek, Lamar, central Pennsylvania: Jour. Sed. Petrology, v. 29, p. 391–401.

Inman, D. L., 1953, Areal and seasonal variations in beach and near-shore sediments at La Jolla, California: Tech. Mem. 34, Beach Erosion Board, Offices of Chief of Engineers.

Kagami, H., 1961, Submarine sediments off Sakata, Yamagata, Japan: Jap. Jour. Geology & Geography, v. 32, p. 397–409.

Kahn, J. S., 1956, Analysis and distribution of packing properties of sand-size sediments, I and II: Jour. Geology, v. 64, p. 385–395, 578–606.

Kolmogorov, A. N., 1941, Über das logarithmisch normale Verteilungsgesetz der Dimensionen der Teilchen bei Zerstückelung: C. R. Acad. Sci., U.S.S.R., v. 31, p. 99–101.

Krumbein, W. C., 1934, Size frequency distribution of sediments: Jour. Sed. Petrology, v. 4, p. 65–67.

——— 1959, Trend analysis of contour-type maps with irregular control-point spacing: Jour. Geophys. Research, v. 65, p. 823–834.

——— 1961, The "sorting out" of geological variables illustrated by regression analysis of factors controlling beach firmness: Jour. Sed. Petrology, v. 29, p. 575–587.

——— and Tukey, J. W., 1956, Multivariate analysis of minerologic, lithologic, and chemical composition of rock bodies: Jour. Sed. Petrology, v. 26, p. 322–337.

Loveland, R. R., and Trivelli, A. P. H., 1927, Mathematical methods of frequency analysis of sizes of particles: Jour. Franklin Institute, v. 204, p. 193–217, 327–389.

McIntyre, D. D., 1959, The hydraulic equivalence and size distributions of some mineral grains from a beach: Jour. Geology, v. 67, p. 278–301.

Miller, R. L., 1956, Trend surfaces, their application to analysis and description of environments of sedimentation: Jour. Geology, v. 64, p. 425–446.

———, and Kahn, J. S., 1962, Statistical analysis in the geological sciences: New York, John Wiley & Sons.

Mumpton, F. A., and Roy, R., 1958, New data on sepiolite and attapulgite, *in* Clays and clay minerals, Natl. Acad. Sci.–Natl. Research Council Pub. 566, p. 136–143.

Nederlof, M. H., 1959, Structure and sedimentology of the Upper Carboniferous of the upper Pisuerga Valleys, Cantabrian Mountains, Spain: Leidse Geologische Mededelingen, v. 24, p. 603–703.

Ōki, J., 1961, Metamorphism in the northern Kisco Range, Nagano Prefecture, Japan: Jap. Jour. Geology & Geography, v. 32, p. 479–506.

Rao, C. R., 1952, Advanced statistical methods in biometric research: New York, John Wiley & Sons.

Rogers, J. J. W., 1960, Geologic interpretation of frequency distributions: Internat. Geol. Cong., 20th, Norden 1960, pt. 21, p. 275–280.

SARMIENTO, R., 1961, Geological factors influencing porosity estimates from velocity logs: Am. Assoc. Petroleum Geologists Bull., v. 45, p. 633–644.

SIMPSON, S. M., 1954, Least squares polynomial fitting to gravitational data and density plotting by digital computers: Geophysics, v. 19, p. 255–269.

STEHLI, F. G., and HOWER, J., 1961, Mineralogy and early diagenesis of carbonate sediments: Jour. Sed. Petrology, v. 31, p. 358–371.

STEVENS, R. E., and NILES, W. W., 1960, Chemical analysis of the granite and diabase, pt. 1 *of* Second report on a cooperative investigation of the composition of two silicate rocks: U.S. Geol. Survey Bull. 1113, p. 3–44.

TAYLOR, S. R., SACHS, M., and CHERRY, R. D., 1961, Studies of tektite composition: Geochim. & Cosmochim. Acta, v. 22, p. 155–168.

WARSHAW, C. M., and ROY, R., 1961, Classification and a scheme for the identification of layer silicates: Geol. Soc. America Bull., v. 72, p. 1455–1490.

WEBER, J. N., and MIDDLETON, G. V., 1961, Geochemistry of the turbidites of the Normanskill and Charny Formations, I and II: Geochim. & Cosmochim. Acta, v. 22, p. 244–288.

WHITTEN, E. H. T., 1961, Quantitative areal modal analysis of granite complexes: Geol. Soc. America Bull., v. 72, p. 1331–1359.

HEINZ A. LOWENSTAM

Biologic Problems Relating to the Composition and Diagenesis of Sediments

THE ACTIVITY of the marine biomass today profoundly affects the physical and chemical processes in the oceans and the composition of the sediments that accumulate on the sea floor. The effects of the biomass on physical processes are illustrated by sediment-binding and reef-building activity and by the formation of baffles that locally modify circulation, salinity, and temperature of the water masses. Calcium carbonate and opaline silica, which constitute a major fraction of the total volume of recent marine sediments, are derived essentially entirely from skeletal remains of marine organisms. Hence, biochemical processes account for the precipitation of these compounds in the present oceans. Prior to the existence of life in the oceans, physicochemical processes were responsible for their precipitation instead. In the early phase of the evolution of life, organisms probably affected the precipitation processes of chemical compounds in the sea only indirectly. Following the diversification of life to include biologic systems that were capable of secreting chemical compounds within and on the surface of their cell complexes, biochemical processes made rapid inroads and in time completely displaced inorganic precipitation of calcium carbonate and opaline silica in the sea. The oldest unquestionable mineralized skeletons of marine organisms, composed of calcium carbonate and opaline silica, are from the early Cambrian. The carbonate precipitates of the planktonic Coccolithophoridae and Globigerinidae, which account for most of the pelagic carbonate oozes today, can be traced back in geologic time to

Professor Lowenstam was born in Siemianowitz in easternmost Germany (an area now included within Poland) and came to the United States in 1937. After receiving his Ph.D. at the University of Chicago in 1939 he was curator of paleontology of the Illinois State Museum and later geologist for the Illinois State Geological Survey. He later joined the faculty of the University of Chicago and, in 1952, became professor of paleoecology at the California Institute of Technology. His principal interests have included paleoecology and geochemistry of marine organisms.

Contribution no. 1166, Division of Geological Sciences, California Institute of Technology.

the mid-Mesozoic. The oldest known diatoms, which contribute a major share of the opaline silica deposits in the open oceans, are found in strata of similar age. This indicates that the beginning of major displacement of inorganic carbonate and silica precipitation by marine organisms on the continental shelves occurred only within the last 10–15 per cent and in deep-sea sediment within the last 5 per cent of the age of the earth. Precise tracing through geologic time of the displacement patterns of inorganic by biologic precipitation processes of carbonates and silica in the oceans depends initially on criteria for the differentiation of the sedimentary particles derived from the two sources. It depends further on representative data of the marine sedimentary rocks of all geologic ages. We tend to forget that our present knowledge of the sedimentary records of pre-Tertiary age is limited to shelf-sea deposits, which constitute only a minor fraction of the total oceanic deposits and are representative of only a very limited range of depositional environments. This poses a major problem when we attempt to consider the history of biologic contributions and effects on the oceans and their accumulating sediments as a whole.

Studies of the history of the marine shelf biota have been concerned in the past with evolutionary changes in skeletal morphology. It has become clear that these morphologic changes are basically surface expressions of biochemical evolutionary changes. We may ask whether these evolutionary changes were also accompanied by changes in kinds of mineral precipitates and of the chemical properties of the minerals. While morphologic changes as related to evolution have been considered a matter of fact, there has been great reluctance to consider evolutionary changes in skeletal mineralogy, even when they have been indicated on grounds of comparative anatomy of fossil skeletons. Failure to establish the relationships of the taxonomic problematica, which will be considered later, can be related, in part at least, to a reluctance to consider mineralogic changes in skeletons as having resulted from evolution. Clarification of this point is important, since in the analysis of the sedimentary rocks we are interested not only in what kinds of minerals were contributed by marine organisms, but also in the changes of the specific contributors to the sedimentary particles in the course of geologic time. If it can be shown that changes in skeletal morphology were accompanied by changes in mineralogy or of mineral composition, we would like to know whether these changes were related entirely to biochemical evolution or to changes in the chemistry of the oceans or to a combination of both factors. Hence we face the added problem whether the chemistry of the oceans has remained the same during the last $5–6 \times 10^8$ years.

Dolomitization and the replacement of high-magnesian calcites and aragonites by low-magnesian calcites are considered to be common features of diagenetic alteration processes in shelf-sea carbonate rocks, as found on continents and insular occurrences. Sediments recovered from the ocean floors, ranging in geologic age as far back as the Eocene, are generally free from cementation. At shallower depths, where aragonite is not dissolved, skeletal remains composed of this mineral are commonly preserved in the sediments. Dolomite constitutes a minor fraction of carbonate rocks of post-Eocene age on land and, as far as reported, appears to be very rare in oceanic sediments of similar geologic ages. Present knowledge of the composition and diagenetic alteration of oceanic sediments is still incomplete, and their time-stratigraphic coverage is still very limited. Reference to the oceanic sediments is made only to emphasize that their diagenesis may be different from that of marine sediments transferred to land. This difference may be related to the fact that the interstitial waters of oceanic sediments remain in contact with the oceanic reservoir, whereas those of the marine sediments accumulating on the continents become isolated from the oceanic reservoir, are altered to brines, and are intermittently replaced by fresh water.

The organic compounds of the soft parts from marine organisms and of the matrices which inclose and sheath the crystals of the skeletons must be included in the study of recent marine sediments and of the sedimentary rocks from the geologic past. It appears that properties of the degradation products from the original organic molecules, as preserved in the sedimentary rocks, are not likely to yield information on questions concerning genetic relationships of organisms, biochemical evolution, and their effects on diagenesis. At this early stage of the inquiry into these questions, however, it seems important to consider quantitatively the organic matrices of recent and fossil skeletons.

The following discussion will be concerned with the status of present knowledge of biologic mineral precipitates—their physical and chemical properties and their organic matrices. Problems indicated by the analyses of the data for each of these subjects will be outlined. The physical and chemical properties of the biologic constituents that become part of recent marine sediments will be considered first.

Recent Biota and Their Sedimentary Records

The recent marine biomass is composed of an enormous diversity of organisms that differ in gross morphology, basic building plans, grades of tissue differentiation, biochemical systems, and habits. The basic building material common to all are carbon-based complex organic molecules. Mineral secretion is widespread but is not uniformly developed throughout the various phyla. Following death, the more complex organic molecules that compose the soft parts of the organisms are commonly degraded to simpler molecules. The decomposition products may retain the physical form of tissue fragments, or they may occur dissociated in the sediments interstitially or adsorbed on mineral surfaces.

A variety of mineral species are known to be precipitated by marine organisms. Many species secrete only one mineral type, and the same mineral may be found among species from different phyla. However, there are species that secrete as many as three different minerals in different parts of their bodies. Mineral secretion is commonly localized in the form of endo- or exoskeletons, and in this case the minerals are usually imbedded in the organic matrices on which they were elaborated. Minerals are also found in the form of isolated spicules inclosed or rooted in the tissues of the organisms. Spicules are generally monomineralic, whereas in skeletons two minerals may be present, and in these species the two minerals usually occupy separate microarchitectural units. The size, shape, and surface geometry of the minerals are species controlled and commonly differ for the same mineral from species to species. The crystal fabric of carbonate skeletons shows characteristic differences in size, shape, and orientation of the crystal between species. In some classes the crystals in the skeletons are arranged in a sequence of layers, and each layer is characterized by a distinct crystal fabric. Mineral precipitation and the mineral species are determined by the biochemistry of the organisms and particularly that of the surrounding tissues on which their organic matrices are elaborated. There is some evidence from a study of carbonate skeletons that the biochemistry of the organisms affects the chemistry of the mineral precipitates. It has been shown that certain ecologic factors, such as the temperature and chemistry of sea water, commonly affect the chemistry of the minerals and their fabrics, and, in some carbonate-secreting forms, the mineral species.

Following death of the organism, the mineralized hard parts are usually freed and separated from the soft parts, which are subject to rapid decomposition, primarily through biochemical attack by other organisms, particularly bacteria. On the average, the mineralized fraction of the organisms is less subject to dissolution or physical and chemical changes by sea water and by physical, biomechanical, and biochemical agents than is the organic fraction. Hence the incidence of survival of the minerals in the process of transfer from the biochemical systems, which caused their precipitation, to the sedimentary environments is greater, and the ratio of minerals to organic matter in the sediments is reversed when compared to the ratio in the living biomass. The enrichment of biologically derived minerals relative to the organic fraction in the sediments is indicated by the common occurrence of mineralized skeletons, either intact or in fragmented form, and by accumulations composed of single mineral grains that were derived from tissue-rooted or imbedded spicules and the disaggregation products of mineralized skeletons. The ultimate state of preservation of mineralized skeletons in the marine sediments—whether they remain physically intact or become fragmented or disaggregated into individual mineral grains—is determined by the physical and chemical properties of the skeletons themselves in relation to the particular factors of the burial environment. The skeletal properties of concern are the mineral species, the relative surface area of the mineral grains, the effective surface areas and the rates of degradation of their particular organic matrices, and the relative surface area and physical bonding relations of the mineral grains in contact with the organic substrates. The factors of the burial environment that must be considered are physical processes, the chemistry of the waters above and below the sediment interface, biomechanical and biochemical processes as related to sediment ingesting organisms and bacteria, and the rate of sedimentation, which determines the length of time the mineralized skeletons are effectively in contact with burial environmental processes.

It is of interest to inquire how far we have advanced in the investigation of the organic compounds and mineral precipitates in living organisms, particularly with reference to their contributions to marine sediments. We wish to know in particular about the different species of minerals that are secreted by members of the marine biomass and the extent to which it becomes possible to differentiate them in the sedimentary occurrences from minerals of inorganic origin. There is the question whether we have advanced sufficiently in the analyses of the clay-sized fraction of the sediments to relate mineral grains to specific biologic sources on the basis of what is known about the morphology of spicular precipitates by organisms. Further, it is important to determine what we have learned about the relations of the physical and chemical properties of the biosynthesis products to the physiology of the organisms and to the ecologic factors that act upon them. Specifically, we must ask the question as to what ways the biologic and ecologic effects are reflected by the physical and chemical properties of the mineral precipitates and the methods that can be applied to differentiate them. It is the purpose in the following to examine the state of progress in each of the areas mentioned, to indicate existing problems and, where possible, methods of approaches that may lead to their solution. Since information on the biologic and ecologic relations to their biologic sources is considerably greater for the mineral precipitates than for their organic compounds, the data on organism-secreted minerals will be considered first.

MINERAL DISTRIBUTION IN BIOLOGIC SYSTEMS

Figure 1 shows the various chemical compounds that are known to be secreted by different marine organisms and their distribution by phylum. The chemical compounds are arranged in the order of their relative abundance as products of secretion by the marine biomass. The numbers inserted under the headings of different phyla signify their estimated importance as secretionary agents of the chemical compounds. Carbonates and silica, as indicated by the pelagic organisms, constitute, in terms of secretionary volumes, the two major chemical compounds

Minerals	Bacteria	Algae	Protozoa	Porifera	Coelenterata	Bryozoa	Brachiopoda	Annelida	Mollusca	Arthropoda	Echinodermata	Hemichordata	Chordata
Carbonates (Ca)	?	√ (1)	√ (1)	√ (3)	√ (2)	√ (2)	√ (3)	√ (2)	√ (1)	√ (1)	√ (2)	√ (3)	√ (3)
Silicates		√ (1)	√ (1)	√ (2)					?				
Phosphates (Ca)							√ (2)		√ (3)	√ (1)			√ (1)
Oxides (Fe)			√ (1)						√ (2)				
Sulfates (Sr, Ba)			√ (1)										

Fig. 1.—Distribution of skeletal mineral types according to phylum. Numbers indicate relative importance of phyla as agents of secretion.

that are precipitated by members of the marine biomass, whereas phosphates, oxides, and sulfates are minor. The data show that carbonate minerals are by far the most widely distributed precipitation products among the biologic systems in the oceans. Carbonate minerals are found among all phyla of the animal kingdom and are also precipitated in the plant kingdom by algae. The question mark under bacteria indicates the possibility that carbonate minerals may also be secreted by marine species of this group. There are numerous references in the literature to such precipitation, but the statements are either vague or uncertain because of controversy on the subject. Dr. Greenfield has recently isolated some marine bacteria that concentrate $CaCO_3$, and the strains supplied to us by him have been induced in sea water to precipitate either calcite or aragonite, depending on the culture methods. It still remains to be determined whether similar precipitates are produced by these organisms in nature, and, if so, under what marine conditions.[1] Compared to the carbonates, silica and phosphates are shown to be considerably more limited in phylum distribution, and the oxides and sulfates even more so. As has been pointed out by several investigators, silica is precipitated by aglae and by the animals representing the lowest grades of organization, whereas phosphates are secreted by species from

	Algae	Protozoa	Porifera	Coelenterata	Bryozoa	Brachiopoda	Annelida	Mollusca	Arthropoda	Echinodermata	Hemichordata	Chordata
Carbonates												
Aragonite	+	+		+	+		+	+	+		+	+
Calcite	+	+	+	+	+	+	+	+	+	+		
Arag. & Calcite			?	+	+		+	+	+			
"Amorphous"				+			+	+				
Silicates												
"Opaline"	+	+	+					?				
Phosphates												
Hydroxyapatite						+						+
Undefined								+	+			
+ Calcite									+			
Oxides												
Magnetite								+				
Goethite								+				
Magn. & Goeth.								+				
Amorphous (Fe)		+						+				
+ Aragonite								+				
Sulfates												
Celestite		+										
Barite		?										

FIG. 2.—Distribution of skeletal mineral species according to phylum

more advanced animal phyla (Vinogradov, 1953). A possible exception to the rule is indicated by the reported development of siliceous spicules in a molluscan species of the Onchidiidae (Labbé, 1933).

Figure 2 shows the status of mineral species identifications, arranged in the same order and in the same phylum sequence as in figure 1.

[1] Since the presentation of the talk on which this paper is based, Greenfield has published an abstract (1963) in which he indicates that the bacteria in his experiments precipitated only aragonite and that he has found in the fine fraction of Florida Bay sediments aragonite-encased bacteria.

Separate columns have been inserted to show where several minerals, and which particular mineral species, have been found to compose different parts of the same skeletons. The data indicate a considerable diversity of biologic mineral precipitates and some complex combinations of minerals within single skeletons of organisms from the marine biomass. Among the carbonates, the minerals calcite and aragonite are shown to be widely distributed among phyla, and these minerals are found in separate parts of single skeletons from species of five and possibly six phyla. Amorphous carbonate precipitates from organisms have been less thoroughly investigated; hence their precise composition and their distribution are still poorly known. Similarly, there are few data on the precise chemical composition of the amorphous opaline silica precipitates, particularly with reference to their states of hydration (Vinogradov, 1953). The same applies to the calcium phosphate minerals, which have been investigated from a few vertebrate skeletons and inarticulate brachiopod shells. The question mark for barite under Protozoa refers to the reported occurrence of minute granules of this mineral in the protoplasm of Xenophyophora. It is not known at present whether these organisms secrete barite or incorporate sedimentary particles of this mineral into their protoplasm (Vinogradov, 1953).

Figure 2 shows that Protozoa and Mollusca precipitate the largest number of different mineral species, whereas the Echinodermata and Hemichordata secrete only one kind of mineral. The distribution patterns of the different mineral precipitates among the phyla seem to be random. Yet, upon closer inspection and despite some exceptions, there appear to be some systematic differences concerning the number of minerals found in single species, as compared to entire phyla, and the numbers and the disposition of several minerals with reference to skeletal development in organisms within groups of phyla. Among the algae and the animal phyla of lowest structural compexity through the cellular grade of tissue construction, only a single type of mineral per species is known to be precipitated, although species of different families, orders, or classes may secrete different minerals.[2] Among the more complex animals from the Radiata through the Schizocoela, there is a tendency for two different minerals to compose different parts of the skeletons, and in some cases still other mineral species are found as precipitates of other anatomical parts in the same organisms. This is particularly true for the Mollusca, where, for example in the Patellacea, the shells of an individual may be composed of aragonite and

[2] Possible exceptions are certain sponges reported to have calcitic and aragonitic spicules and, in one case, layers of calcitic and opaline silica spicules (Vinogradov, 1953). There is some question whether some of these organisms may not belong to other phyla (Hutchinson, *in* Vinogradov, 1953), and there is also the question whether some of their reported minerals are not from epiphytes of extraneous contaminants trapped in their tissues.

calcite and the radular teeth capped by goethite crystals (Lowenstam, 1962*a*). At the highest structural grade, the Enterocoela, the skeletal secretion products are limited to a single mineral, which is the same for all species of an entire phylum. Only in the Pisces, a class of the Chordata, does one find a second mineral precipitate, and this is localized in the otolites, which are not part of the skeleton. It would thus appear from the general distribution pattern of minerals, when considered in combination with the previously noted range of silica and phosphates among the phyla, that there is some relationship between mineral precipitates and phyletic grade.

Figure 2 includes some new data on mineral precipitates by marine organisms that have recently been obtained in our laboratory. The new data refer in the carbonates to the occurrence of aragonite and calcite in the skeletons of certain Hydrozoa and Octocorallia (phylum Coelenterata) and in the skeletons of some Cirripedia (phylum Arthropoda). The occurrence of a phosphate mineral in the Mollusca constitutes the first record of this compound in this phylum and hence extends its phyletic distribution range. We have found that some tectibranch Gastropoda have phosphatic gizzard plates, which contain 12 per cent P_2O_5 (wet chemical analysis). X-ray diffraction patterns show a low degree of crystallinity, and the particular phosphate mineral has not yet been identified.

As to the oxides, goethite has been identified as one of the minerals that occur, together with magnetite, in the capping materials of the radular teeth in some temperate, tropical, and deep-sea chitons (phylum Mollusca). In addition, amorphous ferric oxide minerals have been found to form the cement of the tests of certain arenaceous Foraminifera from the deep sea and of the outer shell layer of at least two pelecypod species taken from depths of 400 meters and 6,000 meters. The inner shell layer of the particular pelecypods is composed of the mineral aragonite. The ferric oxide minerals are still under investigation to determine whether they are hydrated, and the results of their precise composition will be published elsewhere.

The new data extend the range in phylum distribution of certain minerals already known to be biologic precipitates, and they further indicate the presence of previously unreported minerals among biologic secretion products. These data are presented here only to stress how little is still known about the minerals that are synthesized by the component parts of the marine biomass. This is further underlined by the uncertainties noted earlier concerning the precise mineral form of some of the carbonates, silica, and phosphates reported among biologic precipitates. The best example to illustrate this point is the iron oxide minerals. It has been shown quite recently that magnetite is precipitated by chitons and goethite by aspidobranchian gastropods (Lowenstam, 1962*a*, *b*). Until then, iron oxide minerals that had been found in fossil marine sediments were considered to be of inorganic origin, either derived from terrigeneous sources or throught to represent authigenic minerals. The present data indicate that Mollusca and Protozoa precipitate at least three different iron oxide minerals, and they show that the organisms that secrete them are common and range from shallow to deep waters. Hence the iron oxide minerals from recent marine sediments must be re-examined in each occurrence to determine whether they are of inorganic or biologic origin or derived from both sources. Data are needed on the physical and chemical properties of the precipitates from the different sources to determine whether it is possible to distinguish them in mineral occurrences of marine sediments. The problem is not unique to the iron oxide minerals, as indicated by the previously noted uncertainties concerning various biologic mineral precipitates. It appears that, in the past, minerals found only as single grains in marine sediments, lacking unique surface features and not known to have been derived from mineralized skeletons—again as judged from sedimentary occurrences—were automatically considered as of inorganic origin. In other words, we have relied too heavily on the sedimentary occurrences of biologic mineral precipitates instead of determining first which mineral species are precipitated by the various members of the marine biomass. The literature dealing with the anatomy of soft-bodied organisms contains many references to minute granules, spicules, and localized mineral inclusions. Wet chemical analyses have been performed on some, but there are hardly any data on their mineralogy. There is currently increasing interest in the mineralogy from particles of the clay-sized fraction of marine sediments. It is clear that progress in relating the constituent grains to their sources of origin will depend largely upon progress of our knowledge of the minerals precipitated by the living organisms.

MICROARCHITECTURE AND SURFACE GEOMETRY OF BIOLOGIC SEDIMENT PARTICLES

Among the recent marine sediments biogenic carbonate and silica deposits are common, and there are some localized occurrences of enrichment in biogenic phosphates. Aside from these, practically all marine sediments contain varying amounts of a variety of minerals derived from biologic sources. There has been increasing interest in apportioning the total recognizable mineral fraction of marine sediments among the organisms that contributed the different mineral grains and in determining their environmental derivation. Attempts to identify the contributing agents of mineral particles to the sediments have been concerned largely with carbonate-secreting organisms. There are several reasons why this should be so. First, research of recent marine sediments has until recently been concerned largely with shelf deposits, where sedimentary biogenic carbonates are common but biogenic silica is not. Second, the phylum spectrum of carbonate-secreting organisms is considerably greater than for any other biologic mineral secretion. Third, the development of mineralized hard parts, frequently of large dimensions, is common among carbonate-secreting organisms. Although the phosphate-secreting organisms are commonly fully as large as those that secrete carbonate, the diversity in morphology of the hard parts, differentiation in skeletal-forming minerals, and, above all, the diversity of microarchitectural designs found among the carbonate-secreting organisms are unparalleled by all other mineral precipitates by the members of the marine biomass. Studies of recent carbonate deposits have shown that intact or morphologically identifiable skeletal remains, although of prime interest to the taxonomist, constitute on the average only a minor fraction of the total amounts of the constituent grains derived from organisms. More abundant are skeletal fragments in the sediments that lack identifiable surface features, and there are varying amounts of single crystals.

Considerable effort has been made to differentiate microarchitectural units, to define their crystal fabrics, and, in certain groups of organisms, to determine the surface geometry of the crystals that characterize the carbonate skeletons—all as an aid to identification (Bøggild, 1930). These data have become most useful for the identification of biologic contributions of skeletal remains to the coarse-sized fraction (larger than $\frac{1}{8}$ mm.) in which the constituent grains have lost their original characteristic surface features as the result of mechanical or biomechanical abrasion or as the result of biochemical resorption largely through sediment-feeding processes. Much of the subjectivity in identifying the biologic sources of constituent grains by surface inspection under the microscope has been removed by impregnating a representative aliquot of the coarse fraction and examining the microstructures in thin sections (Ginsburg, 1956). There is now a fair amount of published data on the identifications of constituent particles from the coarse fraction of biogenic carbonates, based either on microscopic examination of surface features of the particles or on the thin-section examination, utilizing the microstructure of the grains. Pertinent references are listed in the bibliography by Graf (1960). The data show that the taxonomic ranks conventionally used for groupings of the identifiable biologic fractions differ widely, ranging all the way from the genus to the phylum level. Except for the algal remains of the genus *Halimeda*, most groupings are by orders or higher taxa. This indicates a low degree of resolution in identification levels. Considering the primary aim of most of these studies—to determine the percentages of the major biologic contributions of carbonate particles to the sediments and to assess their rank of importance relative to each other—the conventional broad categories would seem adequate for a first-order approximation. However, most categories are at present taxonomically so broadly defined that we cannot be certain that some significant compositional differences between samples from similar depositional environments are not obscured. Failure to differentiate more precisely sedimentary particles that are currently grouped by classes and phyla is related in part to inadequate data on microarchitectural differences in some groups of organisms. In others, where the microstructure has been extensively investigated, it has been found either that the differences are generally small, as, for example, in the case of the Echinodermata, or that quite similar microarchitectural designs may be found recurring in members of related classes, as is the case in the Pelecypoda and Gastropoda. Hence, it is quite clear that even for the analyses of the biologic contributors to the coarse fraction of sedimentary carbonate there is a need for more precise data on the microarchitecture of the living organisms.

With regard to the silt- and clay-sized carbonates that occur in unsorted sediments or form oozes, reference has been made earlier to the problem of relating particles of carbonate minerals to their biologic sources, since there is still a gap in knowledge of their distribution in the total biomass. The problem concerns the shelf carbonates more than pelagic oozes of the open oceans, since the latter consist largely of uniquely shaped coccolith platelets and *Globigerina* tests or their disaggregation products. The amount of carbonate contributed by benthonic organisms is small. On the shelves, with their highly diversified carbonate biota, the constituent grains of fine-grained carbonates may be derived from many biologic sources. The few published studies of clay-sized shelf carbonates indicate that the percentage of particles that can be assigned with certainty to their derivational sources is quite small (Bramlette, 1926; Ginsburg, 1953; Cloud, 1962). The problem arises from the fact that most of the particles consist of single crystals. As shown in the case of the occurrences of sedimentary aragonite needles, whose origin has been debated for some time, we know of several independent sources that may produce similar single aragonite crystals. Algae belonging to the Codiaceae, Dasycladaceae, Nemalionaceae and Chaetangiaceae precipitate needle-shaped aragonite crystals on their surfaces, and in species of *Halimeda* needles inclose all filaments (Porbeguin, 1954; Lowenstam, 1595; Cloud, 1962). It has been shown that these algae are common among subtropical and tropical shoal-water biota (Lowenstam and Epstein, 1957). In most species the aragonite needles are weakly aggregated, and, after death, as soon as the filaments become decomposed, the needles are released. The skeletons of scleractinian corals and of many pelecypods and gastropods consist also of needle-shaped aragonite crystals held firmly together by organic matrices. Tumbling-barrel experiments, simulating natural physical abrasion processes, show that considerable numbers of single needles can be separated from these biologic skeletons (Chave, 1960). Aragonite needles experimentally precipitated from sea water resemble those precipitated by algae (Gee, Moberg, and Revelle, 1932; Cloud, 1962), and until algal sources for aragonite needles became known, their sedimentary occurrences were generally considered physicochemical precipitates (Vaughan, 1917, 1924; Smith, 1940). Nearly all known occurrences of sedimentary aragonite needles coincide with or are close to areas where algae, corals, and molluscs are common and where conditions deemed necessary for physicochemical precipitation of carbonate do not exist (Lowenstam, 1955; Lowenstam and Epstein, 1957; Chave, 1960). The notable exception is Grand Bahama Bank, west of Andros Island, where algal populations are small and sedimentary aragonite needles are abundant. Data on the range of isotopic ratios for both carbon and oxygen of the sedimentary needles were found to lie within the median range of the algal carbonates, and this has been interpreted to indicate that the sedimentary needles are primarily of organic origin (Lowenstam and Epstein, 1957). A very thorough study of the organisms, water chemistry, and sedimentary composition, based largely on the observation that Ca^{++} and $CO_3^{=}$ ion concentrations decrease from the bank edge toward the site of needle deposition, has led to the strong belief that 75 per cent of the sedimentary needles are products of physicochemical precipitation (Cloud, 1962). Electron microscopy, utilizing the shadowgraphic technique, indicates that sedimentary and algal needles have similar crystal forms, and hence this technique contributes little to the resolution of the problem (Cloud, 1962). The application of the replica technique in electron microscopy (Pl. I, *1*) brings out details of the surface texture of needle fragments of a species of the algal genus *Rhipocephalus* which cannot be seen in shadowgraphic pictures. It may well be that, when algal, coral, mollusc, and supposed physicochemical precipitates of aragonite needles are compared by means of replicas, it may become possible on grounds of their forms and surface features to distinguish between particles from each of these sources.

Another potential source of silt- and clay-sized carbonate particles is dermal spicules, which are usually secreted as single crystals. Judging from the zoölogical literature on the anatomy of living marine organisms, carbonate spicules are secreted by a variety of soft-bodied and skeletal-bearing organisms. Some of these spicular occurrences have been illustrated, and others have been casually mentioned, in the *Treatise of Invertebrate Paleontology*. In contrast to skeletal crystals, spicules show distinct morphology and surface features on the order and family levels and minor differences on the genus and species levels. Hence their identification in silt- and clay-sized carbonates should pose less difficulty than do crystals derived from the disaggregation of carbonate skeletons. It is desirable to have information on all the types of spicular carbonates that are precipitated by recent organisms, on the relative importance of the secreting agents as members of the population where they occur, and on their distribution over the range of environmental conditions that exist in the oceans. The understanding of fossil marine deposits will require a consideration of occurrences of spicules in groups of organisms which comprise minor biotic constituents today but were common in the geologic past.

A few examples from known spicular occurrences in living organisms will suffice to illustrate their distinctive morphology and, hence, to suggest the possibility that their occurrences in sedimentary carbonates may be related to the biologic sources from which they were derived. Plate III, 7, shows a photomicrograph of the aragonitic burr-shaped dermal spicules of a colonial tunicate, of the family Didemnidae, and fragments of single crystals mechanically separated from them. The spicule-secreting tunicates are common in shallow-water environments from the tropics to the arctic. Hence their spicules should be widely distributed in silt- and clay-sized carbonates or in the carbonate fractions of recent sediments. However, tunicate spicules have been mentioned only rarely in analyses of the constituent particles of recent carbonate sediments, and all recorded occurrences refer to the burr-shaped forms of Didemnidae (Bramlette, 1926; Ginsburg, 1953; Cloud, 1962). Echinated spicules have been described from species of some other tunicate families found in shallow water, and branched spicules are known from the branchial sac of some deep water representatives of still another family (Herdman, 1910). To the knowledge of the writer, neither type of spicule has been reported from recent marine sediments.

The Aplacophora, a class usually assigned to the Mollusca, is widely represented from a few tens of fathoms to at least 4,000 fathoms (Heath, 1911, 1918). The organisms are worm shaped, and the epidermis is usually densely covered with aragonitic spicules. Plate II, *2*, shows an individual with the spicules in place, and *3* and *4* show the fragmentary appearance of the separated spicules. Spicules from different species described to date indicate distinctive forms and surface features. As far as the writer is aware, there are no sedimentary records of spicules from Aplacophora.

Another group of organisms that warrant consideration are the Polyplacophora (Mollusca). The girdle of many species is covered by aragonitic scales, and the sutural tuffs of some species consist of aragonitic needle-like spicules. Plate III, *8*, shows a photomicrograph of the sutural tuff spicules and fragmentation products of a species of *Acanthochiton.* Fragments showing little of the tapering forms of the entire spicules may become part of the clay-sized carbonate fraction. Neither scales nor spicules have been reported from carbonate deposits. The application of electron microscopy utilizing the replica technique is again indicated as a means of differentiating spicular fragments in clay-sized carbonates from particles derived from other biologic sources.

Calcareous spicules have also been described from the more primitive Terebratellidae, most Terebratulidae and *Lacazella*, a supposed survivor of the Strophomenacea among the articulate Brachiopoda (Blochmann, 1906, 1908, 1912). The spicules consist of calcite, and their forms range from needle shaped to flat fenestrated plates. The spicules occur inclosed in connective tissue of the mantle, the lophophore, and its cirri. In species of *Terebratulina*, as seen in Plate II, *5*, the number of spicules is quite large. Disaggregated spicules from the lophophore support of the same individual are shown in Plate III, *6*. Spicule-bearing brachiopods are locally common on the shelves and were more widely distributed in Cretaceous and Tertiary shelf deposits. Yet, as far as the writer is aware, no sedimentary records of brachiopod spicules have been reported in the literature. The spicule occurrences among the groups of animals cited, except perhaps for the brachiopods, may constitute, on the average, minor contributions to the silt- and clay-sized carbonate deposits. This has been indicated for tunicate spicules on Grand Bahama Bank (Cloud, 1962). Considering that our present knowledge of the distribution of spicular development and of the form and abundance of spicules among members of the marine biomass is still quite limited, it would appear that not until we know more about them will it be possible to assess their relative contributions to the finer fractions of recent carbonates.

The foregoing considerations have touched on but a few of the problems concerning the identification of carbonate-secreting organisms in sedimentary occurrences. Compared to these, even less is known about the biogenic silica and the depositional records of other biologic mineral precipitates.

ECOLOGIC EFFECTS ON SKELETAL MINERALOGY AND ITS CHEMISTRY

Biogeochemical studies of the carbonate skeletons from recent marine organisms have shown that temperature and chemistry of sea water in which the organisms grow may affect the aragonite-calcite ratio, the Mg and Sr contents, and the O^{18}/O^{16} ratios of the carbonates. Changes in carbonate mineralogy caused by temperature changes may be reflected also in changes of the microarchitecture of the carbonate skeletons. The following discussion will be concerned with the state of progress in this area, the extent to which it is possible to distinguish the effects of temperature and water chemistry on the mineralogy and chemical properties of skeletal materials, and the possible application of approaches similar to those shown as promising in the study of the carbonates to other biologic mineral precipitates.

The relations of the carbonate mineralogy to ecologic factors are considered first. An effect of temperature on the carbonate mineralogy has been clearly demonstrated for species that precipitate calcite and aragonite in their carbonate skeletons (Lowenstam, 1954*a*, *b*, *c*, 1960; Dodd, 1963). The calcareous tubes of most serpulid species (Polychaete worms) show an increase in the aragonite-calcite ratio with elevation in environmental temperatures. In samples taken from waters with large seasonal variations in temperature, the aragonite-calcite ratios of consecutive growth increments show cyclical changes, with higher ratios for summer growth and lower ratios for winter increments (Lowenstam, 1954*c*, 1960). Species that secrete calcite and aragonite in their carbonate skeletons and show a relation of their aragonite-calcite ratios to environmental temperatures are now known from four animal phyla: the Coelenterata, Bryozoa, Annelida, and Mollusca. The relation of carbonate mineralogy to temperature indicates differences between groups of organisms, and hence the aragonite-calcite ratios in different species from the same temperature environment are not the same. In some genera, all species show temperature-related changes in their calcite-aragonite ratios from cold to warm waters; in others, colder-water species show this phenomenon, but warm-water species have skeletons composed entirely of aragonite. Among the latter, the temperatures above which aragonitic skeletons are exclusively secreted may differ according to the species, and there are still others in which a trace of calcite is found only at the minimum temperatures of their ecologic range. A temperature effect is also indicated for the carbonate precipitates of certain green, red, and brown algae and the scleractinian corals. Among the algae, aragonite-precipitating species are limited to subtropical and tropical waters, where the temperatures during the coldest months of the year are higher than 16° C. Algal species in the same families or orders ranging into colder waters do not precipitate carbonate at all on their filaments. The scleractinian corals, which secrete aragonitic skeletons in all temperature regimes, show a major decrease in the number of species in waters below 16° C. during the coldest month of the year. All the reef-building corals that have a high skeletal-carbonate metabolism drop out at this critical minimum temperature. Figure 3 shows diagramatically the major differences that were chosen to illustrate the relations between temperature and the aragonite-calcite ratios for the carbonate precipitates in different taxonomic groups. These differences indicate that the response to temperature by the

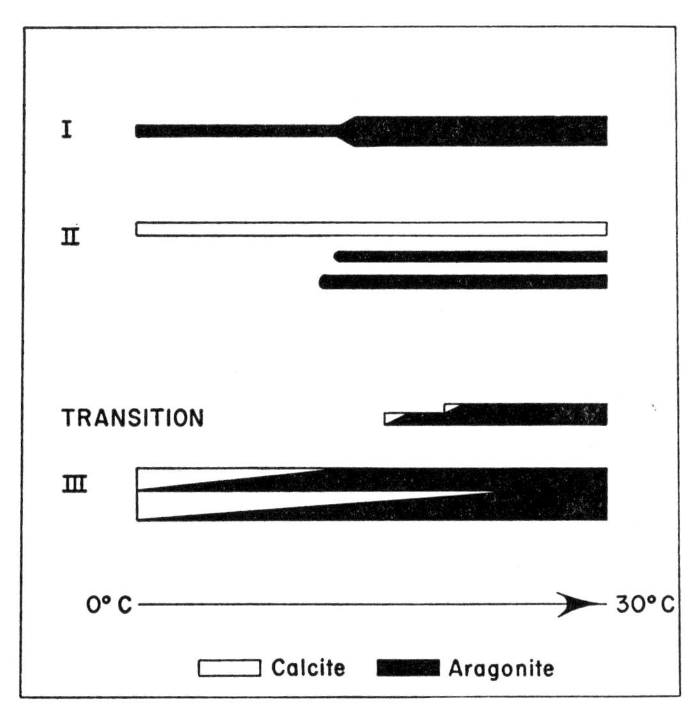

FIG. 3.—Schematic representation of relative stages in complexity of mineralogical response to environmental temperature. Stage I: Animals that secrete only aragonite but in which the number of species is larger in warmer waters. Stage II: Classes or subclasses in which all species of a given order secrete either all aragonite or all calcite, but in which orders that secrete aragonite are confined to warmer water. Transition Stage: Species that differ from those in Stage II by secreting trace amounts of calcite in the colder parts of their ranges. Stage III: Genera with species secreting both calcite and aragonite, with colder-water species secreting relatively more calcite and warmer-water species secreting relatively more aragonite.

aragonite-calcite ratio in the skeletal carbonates is strongly controlled by the biochemistry of the species. It has been recently pointed out that young individuals of the pelecypod, *Mytilus californianus* (i.e., below 15 mm. in length), do not show a temperature effect in their shell mineralogy (Dodd, 1963). This indicates that differences in the physiology at different growth stages for a species must also be considered.

All the species that have been found so far to show an effect of temperature on the skeletal mineralogy of their carbonate skeletons belong to the benthos, and almost all are part of the epifauna and epiflora (i.e., animals and plants living on a substrate). It is widely accepted that the tropical shoal-water biota, best illustrated by those of the coral reefs, have considerably more aragonite-precipitating species and that these tropical species generally have a higher rate of carbonate metabolism than do the biota of the colder waters. Thorson (1950) has shown that the number of shoal-water species of the epifauna increases greatly toward the tropics, whereas the number of infaunal species (i.e., burrowing species) remains about the same from the arctic to the tropics. It was pointed out above that all the aragonite-precipitating species among the algae belong to the epiflora and that they are confined to warm surface waters. Hence, it would appear that the temperature-controlled increase in aragonite over calcite in the biota from the cold to warm surface waters occurs largely, if not entirely, in the epifauna and epiflora. The sedimentary implication of this is that recent carbonate facies composed largely of constituent grains from epibiotic sources in the tropical shoal waters should have, next to reef structures, the highest aragonite-calcite ratios, whereas those from colder-water epibiotic sources should have considerably lower aragonite-calcite ratios. It also follows that shoal-water carbonates or carbonate-rich sediments with infaunal sources for their skeletal carbonates should be similar in their aragonite-calcite ratios regardless of whether they are from tropical or colder waters. Most of their carbonate-secreting organisms are temperature insensitive as far as their aragonite-calcite ratio is concerned, and, at least among the molluscs, most species secrete aragonite only. Hence, we face the problem of estimating the aragonite-calcite ratio of the accumulating sediments from these biologic sources. It will be interesting, after some data have been obtained on this carbonate mineralogy, to compare these with those from pelagic sediments deposited above the depth at which aragonitic skeletons of the pteropods and *Spirula* shells are dissolved.

It has been shown that salinity also affects the aragonite-calcite ratio of skeletons from polychaete worms and certain pelecypods (Lowenstam, 1954*b*, *c*; Dodd, 1963). The aragonite-calcite ratio increases with decrease in salinity. It is not yet clear whether changes in ion concentrations alone are responsible for the observed changes in aragonite-calcite ratios, which would imply solely a physiologic effect, or whether the differences in "salinity" also involve changes in ionic ratios that are related to differences in the chemistry of fresh waters that are mixed with mean ocean waters.

Thin-section studies of the shells from cold- and warm-water species of a few pelecypod genera indicate that changes in the carbonate mineralogy as related to temperature are accompanied by changes in the microstructure (Lowenstam, 1954*c*). The microarchitectural changes are most clearly marked in shells of those cold-water species which consist of an outer calcitic and an inner aragonitic layer, whereas in related warm-water species the two shell layers are composed equally of aragonite. Plate IV, *9* and *10*, shows photomicrographs of longitudinal thin sections through the shells of a cold- and warm-water species of the genus *Chama*. The inner aragonitic shell parts of the two species are similar and consist, according to the nomenclature of Bøggild (1930), of alterations of crossed lamellar and prismatic structural units. The outer shell part of the cold-water species *pellucida* (Pl. IV, *10*), which consists of calcite, shows an inclined prismatic structure, whereas the warm-water species *iostoma* (Pl. IV, *9*), in which this shell part is composed of aragonite, has a crossed lamellar structure and hence is fairly similar to the inner shell parts of both species.

A preliminary survey of the iron oxide minerals from the denticles of the radular teeth in shoal-water chitons indicates that in cold-water species the denticles are composed entirely of magnetite, whereas in temperature- and warm-water species several at present unidentified oxide minerals are present in addition to magnetite (Lowenstam, 1962*b*). It remains to be determined whether the mineralogic changes are related to environmental factors, such as temperature, or to genetically controlled differences of the biochemistry of the species. More data are needed on the mineralogy of the biologic phosphatic and siliceous precipitates before it will be possible to consider possible relations of these minerals to ecologic effects. There are also indications that some second-order effects of factors at present undetermined influence the aragonite-calcite ratios of temperature- and salinity-sensitive species (Lowenstam, 1954*b;* Dodd, 1963).

It has been shown that the Mg content of skeletal carbonates is in solid solution (Chave, 1952; Goldsmith *et al.*, 1955) and that the Mg content of carbonate skeletons is dependent on mineralogy, biochemistry of the species, and environmental temperature. It may also be affected by the salinity of the waters in which the organisms grow (Clarke and Wheeler, 1922; Chave, 1954; Blackmon and Todd, 1959; Pilkey and Hower, 1960; Lowenstam, 1961). The Mg contents generally show a positive correlation with elevation in temperature. The Mg contents of randomly selected calcite-secreting species selected from different temperature regimes on the order or class level all show a temperature effect. The relative concentrations of Mg and the slopes of the Mg-temperature curves differ for different orders and classes, and the slopes of the curves tend to become lower as the grade of structural complexity becomes higher (Chave, 1954). The slopes of Mg-temperature curves obtained from samples of a single echinoid and an inarticulate brachiopod species indicate marked deviations from those for the respective class (Pilkey and Hower, 1960; Lowenstam, 1961). The Mg contents determined for samples of species within some genera of the Foraminifera show little variations with temperature (Blackmon and Todd, 1959). It is not known whether in these Foraminifera there is no temperature effect on the uptake of Mg in these calcitic tests or whether changes in Mg contents with elevation in temperature are so small that they are within the range of the average deviations of the Mg determinations. The relation between temperature and Mg contents in aragonitic skeletons has been investigated only in the scleractinian corals, and the data are inconclusive (Chave, 1954). Clearly, more data are needed to determine whether the Mg contents of biologically precipitated aragonites are sensitive to temperature.

The relation of salinity to Mg contents in carbonate skeletons has been investigated in the calcitic skeleton of a species of echinoid and in some articulate brachiopods (Pilkey and Hower, 1960; Lowenstam, 1961). These data indicate that there is a positive correlation in Mg contents with salinity. In the case of the brachiopods, samples from hypersaline waters have a noticeably higher Mg content than do those from mean ocean salinities at the same temperature. Determinations of the Mg/Ca ratios of the waters from which the biologic samples were collected do not show any difference in ratios. This would tend to suggest that the higher Mg contents of the brachiopod samples from hypersaline waters, as compared to those from mean ocean salinities, are related to the effects of greater ion concentrations on the physiology of the organisms (Lowenstam, 1961).

Turning to the Sr contents of the carbonate skeletons from marine organisms, the factors affecting the concentrations in the carbonate crystals are the same as those known to control the uptake of Mg. The relations between temperature and Sr uptake have been investigated in the low-strontian calcitic skeletons of a species of irregular echinoid and in the articulate brachiopods (Pilkey and Hower, 1960; Lowenstam, 1961). In echinoids the Sr contents are shown to decrease with elevation in temperature, whereas in the articulate brachiopods there is an increase. The slope of the Sr-temperature curve for samples for a single species was found to differ from that based on randomly selected species for the class as a whole.

In the echinoid species examined for Sr-temperature relation, changes in salinity do not seem to affect the Sr contents of their skeletal carbonates (Pilkey and Hower, 1960). However, in the articulate brachiopods differences in salinity are marked by changes in Sr contents of their calcitic shells (Lowenstam, 1961). The Sr/Ca ratios of water samples taken over the bottoms where the brachiopods lived show differences for the hyposaline and hypersaline waters relative to those from mean ocean salinities. The Sr/Ca ratios in the shells for the same temperature are proportional to the Sr/Ca ratio sea water in which the organisms grow. This is in agreement with the experimental growth studies of fresh-water gastropods in waters of different Sr/Ca ratios (Odum, 1951). Again, more data are needed on the relations of Sr contents to temperature, salinity, and water chemistry in the skeletal carbonates from different phyla.

The O^{18}/O^{16} ratios of carbonates laid down in isotopic equilibrium with the surrounding sea water are, unlike Sr and Mg contents, dependent upon temperature and are independent of the carbonate mineralogy and the biochemistry of the species (Urey *et al.*, 1951). Investigations of the O^{18} contents of marine waters indicate that the O^{18} content of waters from the well-mixed oceanic reservoir show little variation, whereas sea water diluted with fresh water, depending on its source, may be enriched in O^{16} to varying degrees. Hyposaline waters are enriched in O^{18} proportional to the degree of insolation relative to that in mean ocean water (Epstein and Mayeda, 1953). Average temperatures calculated from the water-corrected O^{18}/O^{16} ratios of the carbonate from samples of different species at Bermuda show a high incidence of deviations from the mean annual temperature of the local waters (Epstein and Lowenstam, 1953). Data on the temperature distribution obtained from the O^{18} contents of consecutive growth increments from shells of some species of Bermuda molluscs indicate that the shell carbonate is laid down by some species the year round, whereas in other species it is limited to temperature ranges that cover only part of the yearly amplitude (Epstein and Lowenstam, 1953). The amounts of carbonate secreted by organisms at different temperatures within the range of temperatures that they tolerate vary according to the species. The factors indicated to control the skeletal carbonate secretion by different species account largely for the differences in isotopically determined temperatures.

It has been noted that the mineralogy, Sr and Mg contents, and the O^{18}/O^{16} ratios may be affected simultaneously by temperature and the chemistry of sea water, and in some cases also by salinity. Considering possible applications to fossil carbonate skeletons, it has been realized that it would prove difficult to distinguish between the effects of the different ecologic factors when the mineralogy and the various chemical properties are considered separately. Also, in fossil carbonates there is the added problem of distinguishing between original and diagenetically altered minerals and their chemical properties. To determine whether it is possible to distinguish between the effects of temperature and salinity or water chemistry, the Sr and Mg contents were compared with the O^{18}/O^{16} ratios in single shells of articulate brachiopods that were collected from waters of temperatures from 10°–26° C. and of salinities from 30‰–38‰ (parts per thousand). Water samples taken at the living sites of the brachiopods were analyzed for their O^{18}, Sr, Mg, and Ca contents (Lowenstam, 1961). Figures 4 and 5 show plots of the $MgCO_3$ and $SrCO_3$ contents against the water-corrected O^{18}/O^{16} ratios for the samples. The experimental points for samples from insolated waters (salinity = 38‰) are shown to lie clearly outside and above the area defined by samples from waters within ±5 per cent of mean ocean salinity (33.5‰–36.5‰), whereas those from hyposaline waters (salinity = 30‰) are located within the area or barely below. The Sr/Ca ratios determined for the hypo- and hypersaline waters were found to be from 7 to 15 per cent higher than the average of the Sr/Ca ratios for samples to mean ocean salinities, whereas the Mg/Ca ratios of all samples were found to be the same. In figures 6 and 7, the data are treated as in the case of fossil samples, that is, the O^{18}/O^{16} ratios are uncorrected for the O^{18} contents of the waters in which the shells were precipitated. The plots show that the experimental points for samples from hypersaline waters are located above, whereas those from hyposaline waters are below the area defined by the experimental points for samples from ±5 per cent of mean ocean salinity. Samples from the slightly insolated waters of Bermuda and Barbados, which have slightly higher values in O^{18} contents as compared to mean ocean waters, also occupy a position indicating a derivation from hypersaline waters. The data indicate that by means of this method of comparison of different properties it is possible to distinguish in skeletal carbonates between the effects of temperature and water chemistry.

It has been suggested that by comparing the O^{18}/O^{16} ratio of carbonate and phosphate skeletons laid down in isotopic equilibrium with the surrounding water it should be possible to distinguish between temperature and the O^{18} contents of the waters (Urey *et al.*, 1951). A method for extracting reproducible yields of O^{18} from phosphates has been developed, but it is not yet known whether the O^{18}/O^{16} ratios of the phosphates precipitated by marine organisms are in isotopic equilibrium with the O^{18} content of their surrounding waters (Tudge, 1960). The investigation of this question is important in the consideration of other independent approaches for differentiation between the effects of temperature and water chemistry. It is also possible that the chemistry of siliceous skeletons may be utilized in this respect. In the case of the Sr and Mg contents of carbonate skeletons, we should like to determine more precisely their relations to temperature and water chemistry, as distinct from possible effects of salinity. This can be readily accom-

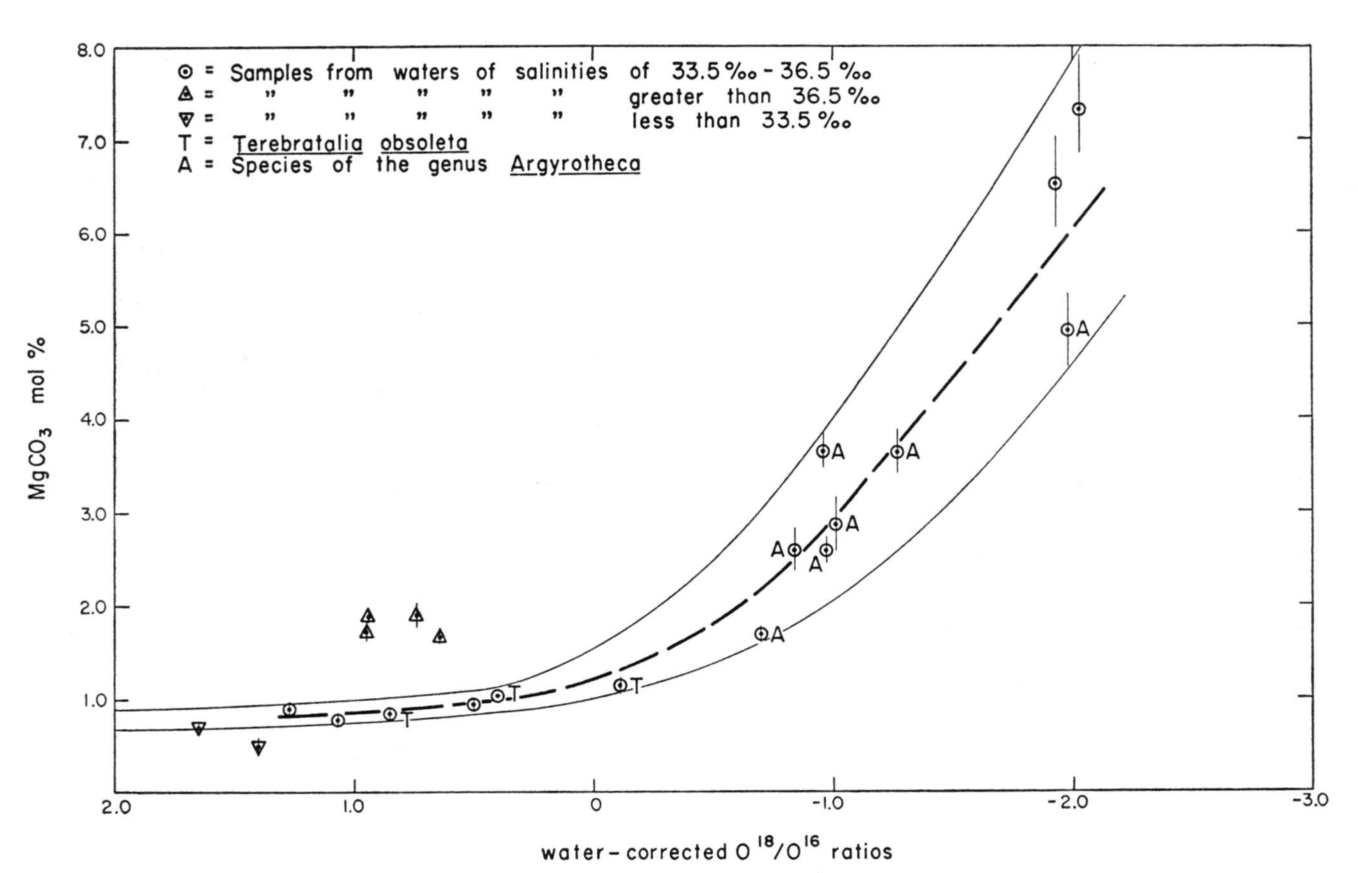

FIG. 4.—Plot of $MgCO_3$ content against O^{18}/O^{16} ratios for recent samples. Data corrected for O^{18} content of water

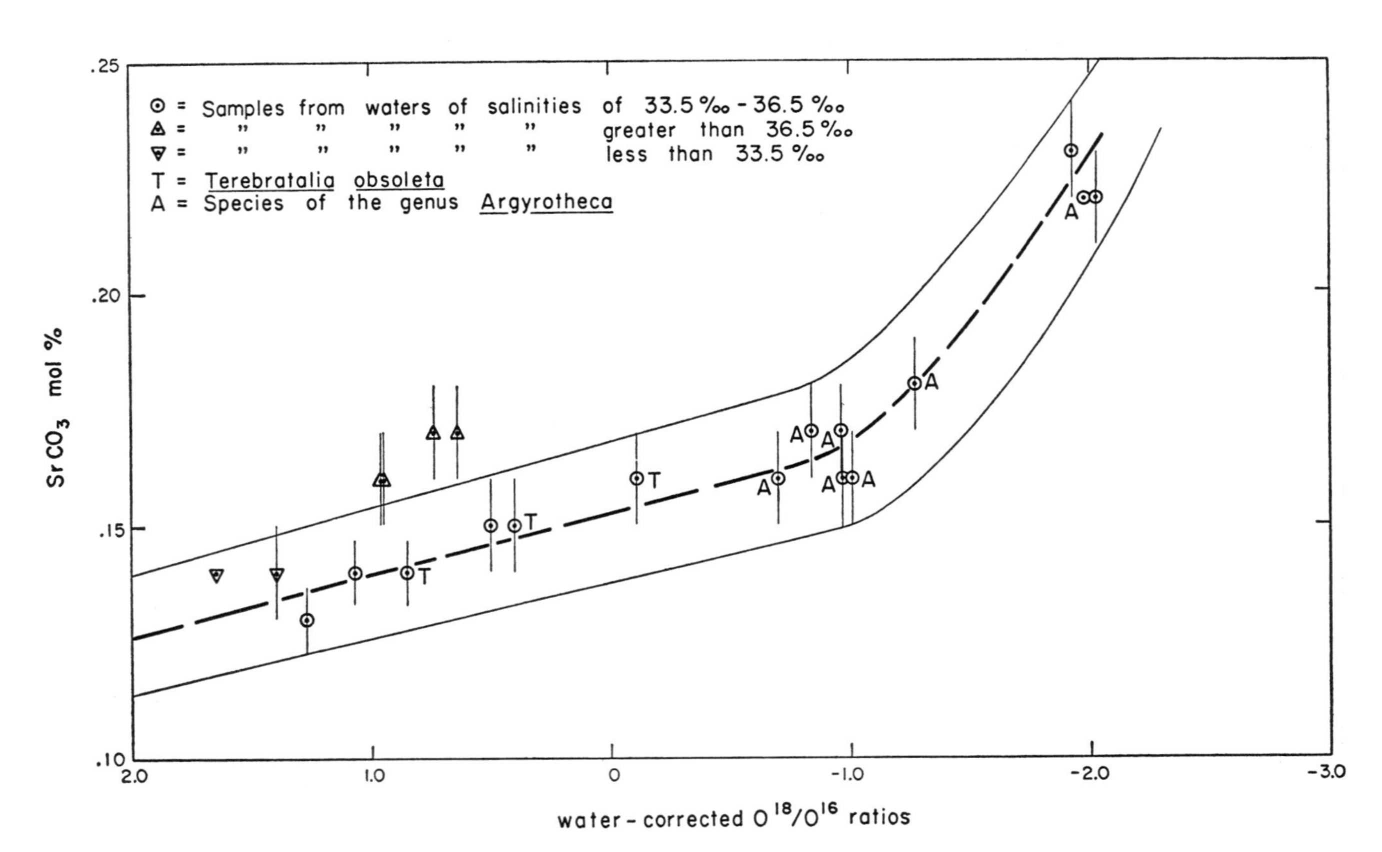

FIG. 5.—Plot of $SrCO_3$ content against O^{18}/O^{16} ratios for recent samples. Data corrected for O^{18} content of water

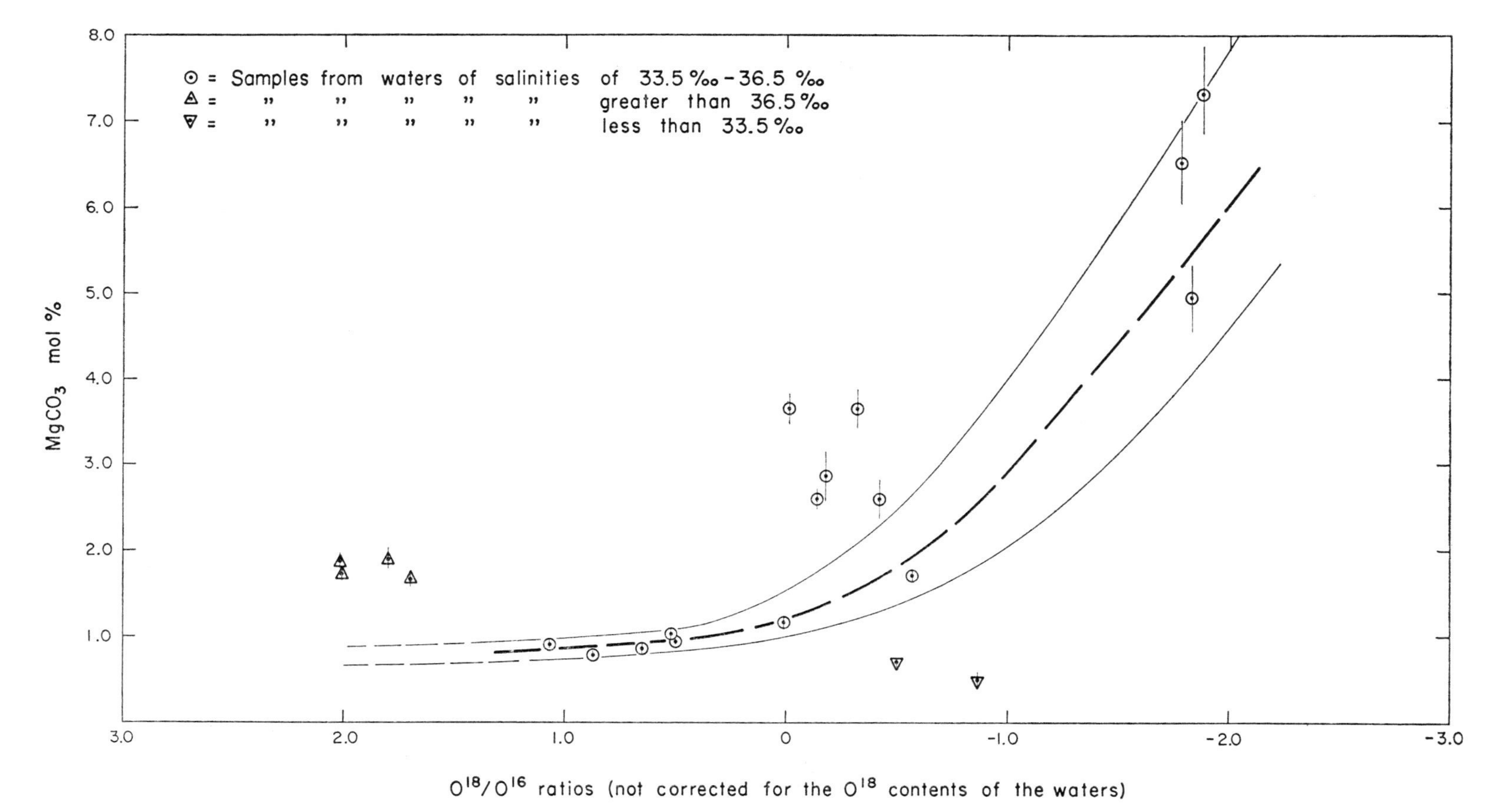

FIG. 6.—Same as figure 4, but data not corrected for O^{18} content of water

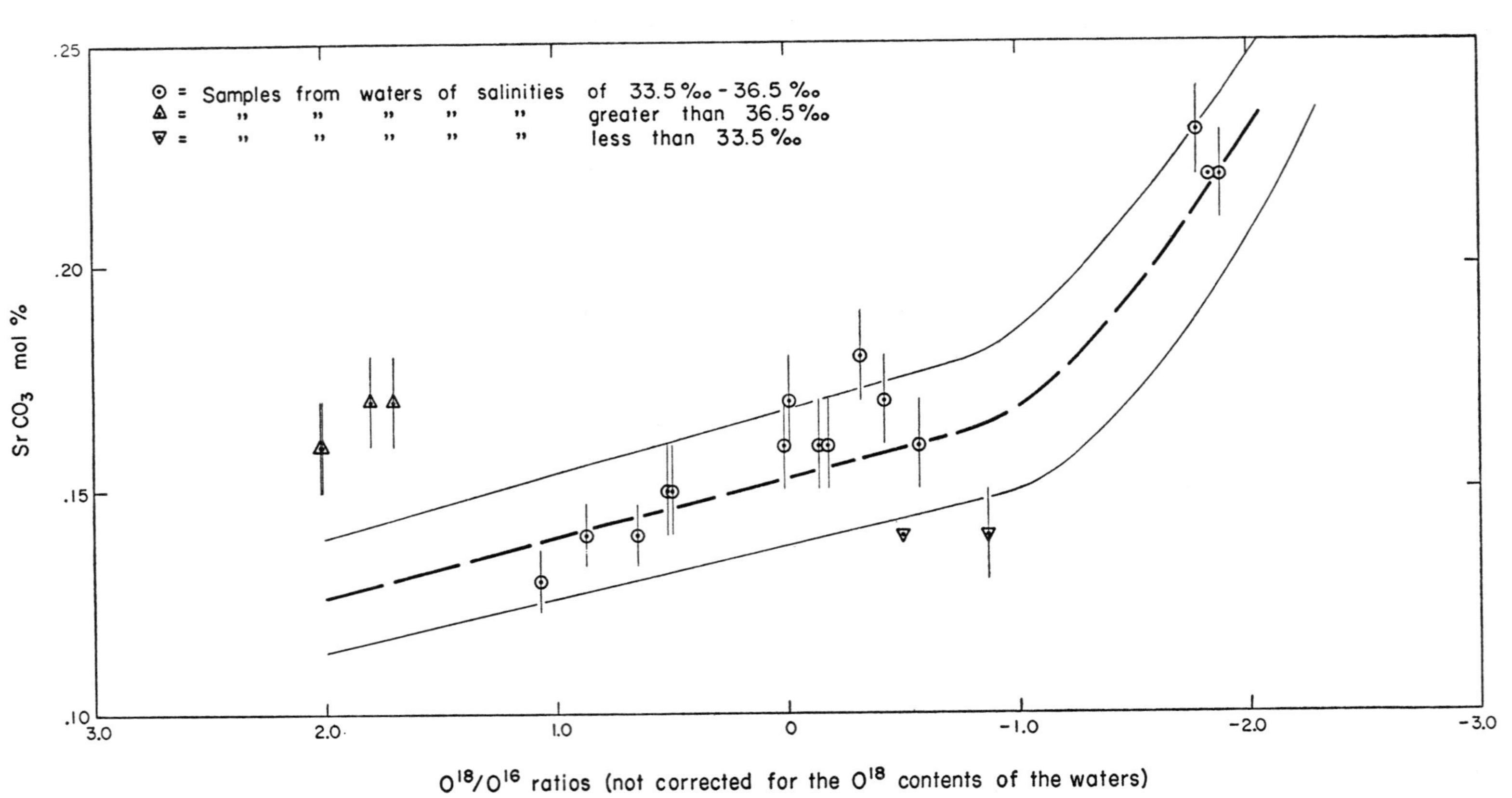

FIG. 7.—Same as figure 5, but data not corrected for O^{18} content of water

plished by experimental growth studies of organisms under controlled laboratory conditions, where all but one of the variables are kept constant for a time. These studies should also determine whether there are at present still other undetected ecologic factors that affect the uptake of these trace constituents by skeletal carbonates. It will be most helpful to investigate the minor trace elements in carbonates, as well as in other chemical compounds precipitated by marine organisms, to determine their relations to ecologic factors. Temperature, salinity, and water chemistry of the marine surface waters over the shelves are subject to considerable fluctuations during the life spans of most organisms. Most of them are seasonal, but some are not. We should like to examine these in as much detail as possible, where they are recorded by the mineralogy, crystal morphology, and mineral chemistry in the skeletons of marine organisms. Preliminary studies utilizing the electron microscope and electron microprobe indicate that this is feasible, and data obtained by means of these instruments should aid in bringing into focus details vital to the understanding of the ecology of the organisms and indicate their paleoecologic and sedimentary implications.

BIOCHEMICAL CONTROL OF Sr AND Mg CONTENTS OF SKELETAL CARBONATES

Biochemistry of organisms has been shown to constitute one of several factors determining the concentration levels of Sr and Mg in the carbonate crystals of skeletal precipitates. Evidence in support of a biologic effect has been derived from the observation that there are differences in the Sr and Mg contents of the skeletal carbonates from species of different classes and, in some cases, between those of different genera within an order (Chave, 1954; Odum, 1957; Blackmon and Todd, 1959). From a study largely based on high-magnesian calcites Chave (1954) concluded not only that there is biologic partitioning of the Mg contents in the calcites on the order and class level but that the biochemistry of the organisms also affects the relation of temperature to the uptake of Mg in the skeletal calcites in a systematic manner. Data indicating biologic control on Sr uptake have largely come from the extensive surveys of the distribution of this element in the carbonate skeletons of a variety of marine organisms (Thompson and Chow, 1955; Odum, 1957). Odum (1957) devised an elaborate scheme based on the relative effectiveness of fluid exchange with the external medium, nucleation sites of the carbonate in relation to tissues, food, and grades of tissue differentiation to account for the apparent randomness of Sr contents in the carbonate skeletons within various higher taxa. Some controversy has developed concerning the magnitude of the biochemical effect on the uptake of Sr and Mg in the skeletal carbonates as compared with the effect of the crystal chemistry and of ecologic factors. The controversy stems largely from the observation that within pelecypod and gastropod genera the differences in Mg and Sr contents of calcitic and aragonitic shells and in samples from different temperature environments can be very small as compared to differences found between genera and classes (Turekian and Armstrong, 1960). This has been interpreted to indicate that the biochemistry on the genus level is the most important factor in determining the concentration levels of Sr and Mg in the carbonate skeletons, whereas from studies largely of other biologic groups, mineralogy has been shown to constitute the most important factor (Chave, 1954; Lowenstam, 1954*a*; Odum, 1957). The clarification of this question is important from the biologic point of view and also when one considers the possibility that biochemical evolution may have effected changes in uptake of Sr and Mg that would be monitored by changes in concentration levels of the skeletal carbonates in fossils. With the notable exception of the studies by Chave (1954) and Blackmon and Todd (1959) on magnesian calcites, the effect of the biochemistry as distinct from that of the mineralogy and ecology is impossible to assess from data in the literature. The trace-element analyses usually lack data on either the mineralogy of the samples, the ecology of the organisms, or both. To define biochemically determined differences between orders, classes, and phyla of the Sr and Mg contents in the skeletal carbonates, it would seem expedient to select samples from species with the same skeletal mineralogy living under identical environmental conditions. In this case the mineralogic and ecologic effects on the carbonate precipitates would be the same and the differences in trace-element concentrations between samples of different higher taxonomic rank would be related to the biochemistry of the organisms. Samples from species of all phyla that were found to secrete aragonite among the shoal-water plants and animals of Bermuda were selected for the investigation of the Mg and Sr contents of their aragonitic precipitation products. Figures 8 and 9 show the distribution of the $MgCO_3$ and of the $SrCO_3$ contents for samples of species within orders, classes, or phyla, depending upon the number of analyses per phylum. Partitioning on the taxonomic levels differentiated here is shown to be widespread but not uniformly recognizable. In the plot of the $MgCO_3$ contents (fig. 8) the data for the algae show an exceptionally large scatter of experimental points with nearly overlapping ranges for the Chlorophyta and Rhodophyta. Two analyses of Phaeophyta show $MgCO_3$ contents that are higher than those of the two other algal groups. The data for the animal samples show a trend of systematically decreasing $MgCO_3$ contents

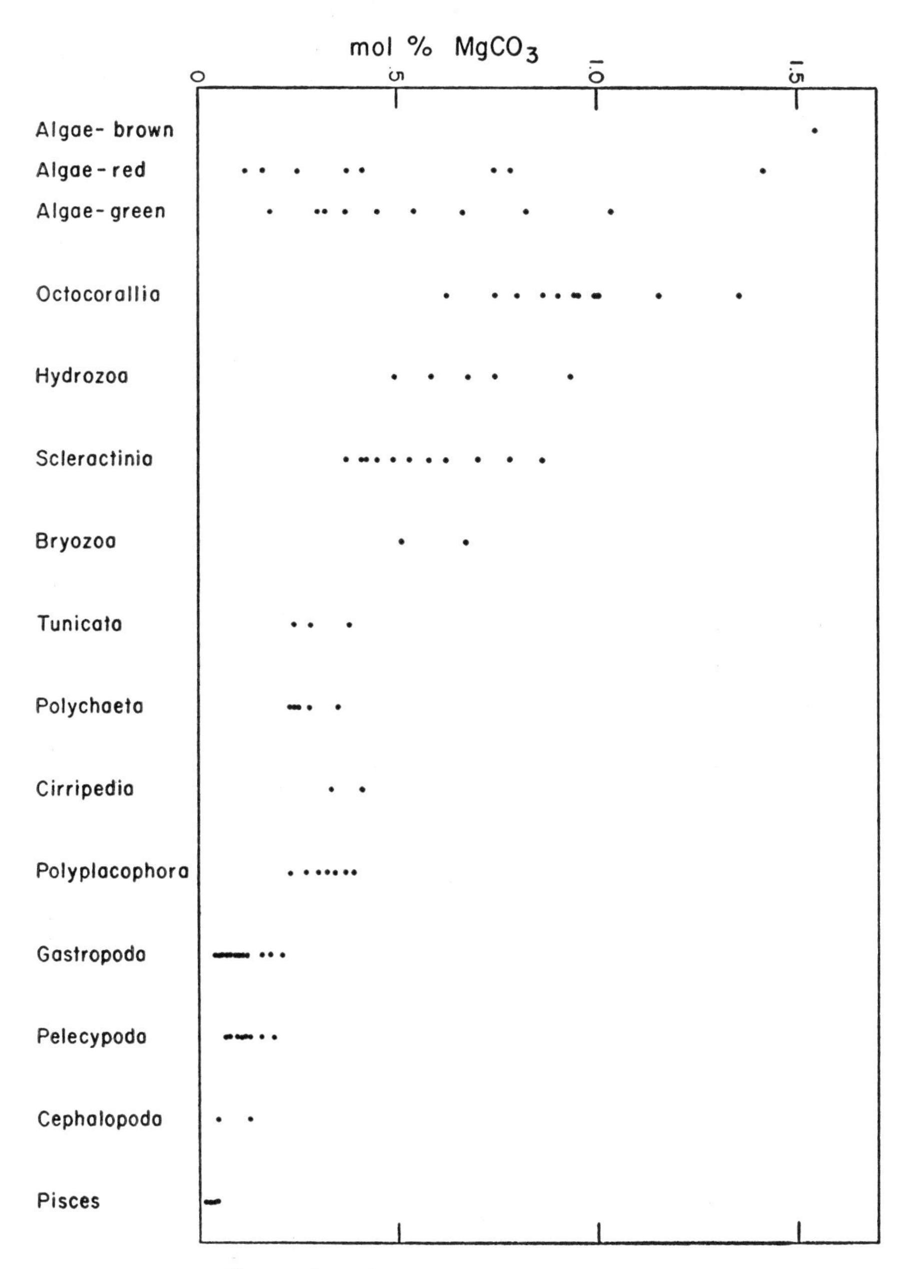

FIG. 8.—Plot of $MgCO_3$ content against phyletic rank

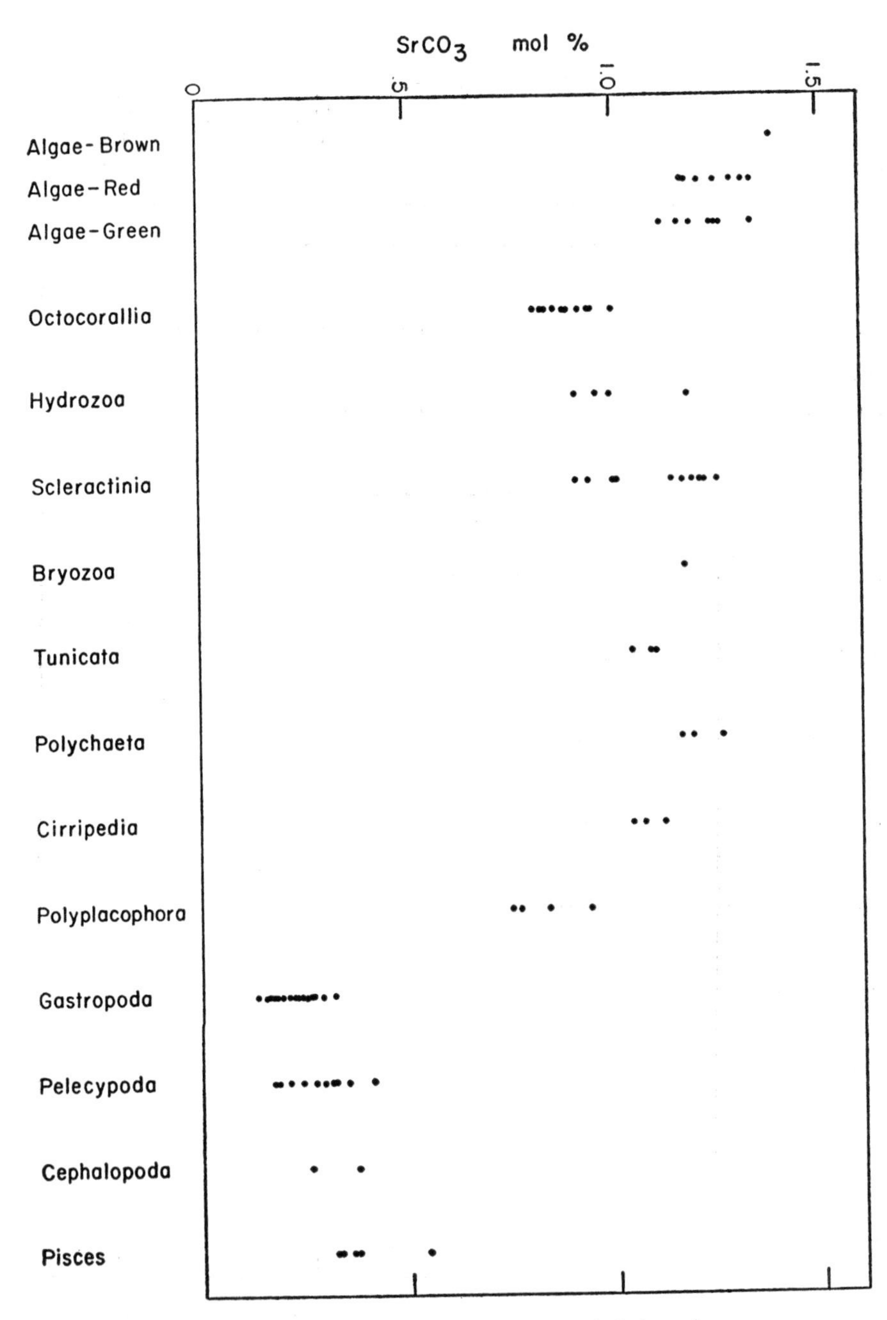

FIG. 9.—Plot of $SrCO_3$ content against phyletic rank

with increase in levels of relative complexity.[3] As to the $SrCO_3$ contents (fig. 9), the algal data show a relation for the Phaeophyta, Rhodophyta, and Chlorophyta similar to that indicated for their $MgCO_3$ contents. The data for the animal samples show a well-defined differentiation into high- and low-$SrCO_3$-bearing groups of organisms. High $SrCO_3$ contents, similar to those found in the algae, are shown to encompass all levels of phyletic grades from the Coelenterata through the polyplacophoran Mollusca, whereas the more advanced Mollusca and Pisces show low $SrCO_3$ contents. There is no overlap in concentration ranges between the high- and low-strontian groups, and the line of demaraction between the two is, curiously, found between two classes within a single phylum—the Mollusca. The fact that organisms with low-strontian aragonites are represented only among the more advanced Mollusca and the Pisces reflects some tendency toward decrease in Sr uptake with increase in phyletic grade. However, instead of a trend of progressive decrease in concentrations, as shown by the $MgCO_3$ contents, there is a sharply defined shift from a plateau of high to one of low $SrCO_3$ contents. Sr and Mg constitute the two major trace constituents that substitute for calcium ions in the aragonitic skeletal carbonates. The tendency for both to decrease as substituting ions for Ca from the less complex to the more highly differentiated tissue grades indicates an increase in power of discrimination by more advanced biochemical systems against ions from the external aqueous medium. The observed result is an increase in purity of the calcium carbonate precipitated by the higher organisms.

The Sr and Mg determinations shown in the graphs include the first data on the aragonitic precipitates of Rhodophyta, Octocorallia, Bryozoa, Tunicata, and Cirripedia. Hence, they contribute to the trace element characterization of skeletally precipitated aragonites that end up as a part of the biologic carbonate fraction in the marine sediments. Considering their implications, the new data on the $MgCO_3$ contents do not alter our previous concept that the $MgCO_3$ contents of biologically precipitated aragonites are generally low, that is, about 1 mole per cent or less. This is borne out by the distribution of magnesium concentrations shown in figure 8, where the highest value, for the Phaeophyta, is 1.4 mole per cent $MgCO_3$. Turning to the $SrCO_3$ determinations, the new data show that high-strontian aragonites, instead of being limited to very few biologic groups, as thought in the past, are actually more widely distributed among the phyla than are low-strontian aragonites. There is by now a total of nine phyla whose aragonitic precipitates have

[3] The only notable exceptions are the tunicates, which on grounds of their embryology are placed taxonomically at a higher level than is indicated by their position on the Mg scale in figure 8.

high $SrCO_3$ contents, ranging from 0.8 to 1.4 mole per cent, as contrasted to two phyla where low $SrCO_3$ contents (i.e., between 0.2 and 0.5 mole per cent) are found.

The survey of the aragonitic precipitates from the Bermuda biota covers only a limited segment of the biologic systems that secrete $CaCO_3$ in the sea. It becomes quite clear that considerably more information is needed on the Sr and Mg contents of biologic carbonates before one can define their distributional abundances, their partitioning by biologic systems, and their differentiation from aragonites that are precipitated physicochemically from sea water.

PHYSICAL AND CHEMICAL PROPERTIES OF SKELETAL ORGANIC MATRICES

Investigations of mineralized skeletons were in the past preoccupied with mineralogy. However, more recently attention has turned to a consideration of the organic matrices of the minerals and the structural organic units, where these are developed. Detailed studies of physical and chemical properties have been made so far only on carbonate skeletons of marine organisms. Electron microscope studies of the decalcified matrices from shells of recent gastropods, pelecypods, and cephalopods have shown that there are major systematic differences in the structure of the lacy network of the organic membranes, and minor ones on the species level (Grégoire, Duchateau and Florkin, 1955). Quantitative studies of the amino acid composition from the decalcified organic matrices have been carried out on the shells of the two West Coast species of *Mytilus* (Hare, 1961, 1963). These studies show that the amino acid composition of the proteins composing the periostracum, ligament, and matrices of the calcitic and aragonitic parts in individual shells differ significantly from one another. In *M. californianus* the protein matrix of the outer calcitic layer was found to be higher in the ratio of acidic to basic residues than that of the inner aragonitic layer. The structural units of the shell that are not calcified have lower ratios of acidic to basic residues than do the calcified ones. The protein matrix of the outer calcitic shell layer of *M. edulis* was shown to contain significantly more glutamic acid residues than that of *M. californianus*, collected from the same habitat. Samples of *M. californianus* from different temperature environments were found to differ systematically in the amino acid contents of their aragonitic matrices.

These data show that a definite beginning has been made to decode biochemical and ecologic information monitored by the physical and chemical properties of the organic parts of mineralized skeletons. Not only do these data add to similar mineralogical information, but, what is more important as more data will become available, they should provide basic information on the chemical and physical processes that determine the precipitation of the particular minerals in different parts of the skeletons and in different biochemical systems. Complex differentiation of proteins is indicated by the amino acid compositions of the different structural components in the *Mytilus* shells. Similar studies on the mineralized skeletons of all sediment-contributing organisms are needed before it will be possible to evaluate the degradation states of their protein matrices in the depositional environments.

Phyletic Distribution of Mineralized Skeletons in the Geologic Past

The evolution of marine life, as documented by skeletal remains in the sedimentary rocks, is marked by sudden appearance, rapid diversification, decline, and extinction of various types of skeletal-bearing organisms. It is of interest to consider, for each of the various chemical compounds that are precipitated by members of the recent biomass, how far back in geologic time we can trace their biosynthesis history, and to examine the compositional changes of the biologic agents that precipitated them in the course of their evolution. Such information should be useful for the determination of changes in kind and amount of minerals contributed by the evolving marine biomass and also changes in major and minor biologic contributions of various minerals to the marine sediments in the geologic past.

Figure 10 shows the silica- and phosphate-secreting organisms by phylum, and their known time-stratigraphic ranges. The width of the bars indicates the estimated importance of the precipitating organisms as compared with other groups secreting the same minerals. The silica-secreting organisms of the Paleozoic and early Mesozoic are represented by species of two of the major extant groups, the Radiolaria and the siliceous Porifera. Among the algae the diatoms, which constitute today the third major element of biologic silica secretion, make their first appearance in the fossil records during Liassic time. However, diatoms became abundant in marine sedimentary rocks only in the late Cretaceous. The mid-to-late Mesozoic also shows a phase of moderate expansion of the silica-precipitating Protozoa with the appearance of the Silicoflagellidae and possibly of certain silica-precipitating Foraminifera. There is still some question about the siliceous tests of the Silicinidae and the siliceous cements reported from the arenaceous tests of *Silicosigmoilina* and *Silicotextulina* in fossil deposits—namely, whether they are original or diagenetic replacement products. The mid-Tertiary occurrences of the Ebriidae indicate a further, temporary expansion of the silica-precipitating plankton. The Ebriidae are with some doubt referred to a distinct element of the Flagellata (Mortet, 1943). There are at present no fossil marine records of the protozoan Heliozoa, which have recent silica-precipitating species.

These data indicate that biosynthesis of silica, producing preservable remains, was carried out from the beginning of the Paleozoic to the mid-Mesozoic by species of only two of the phyla engaged in this process today. A phase of major diversification in silica-secreting organisms on the phylum, class, and order levels occurred in the mid-Mesozoic, but the newly added groups did not become major silica-extracting agents, similar to present forms, until very late in Mesozoic

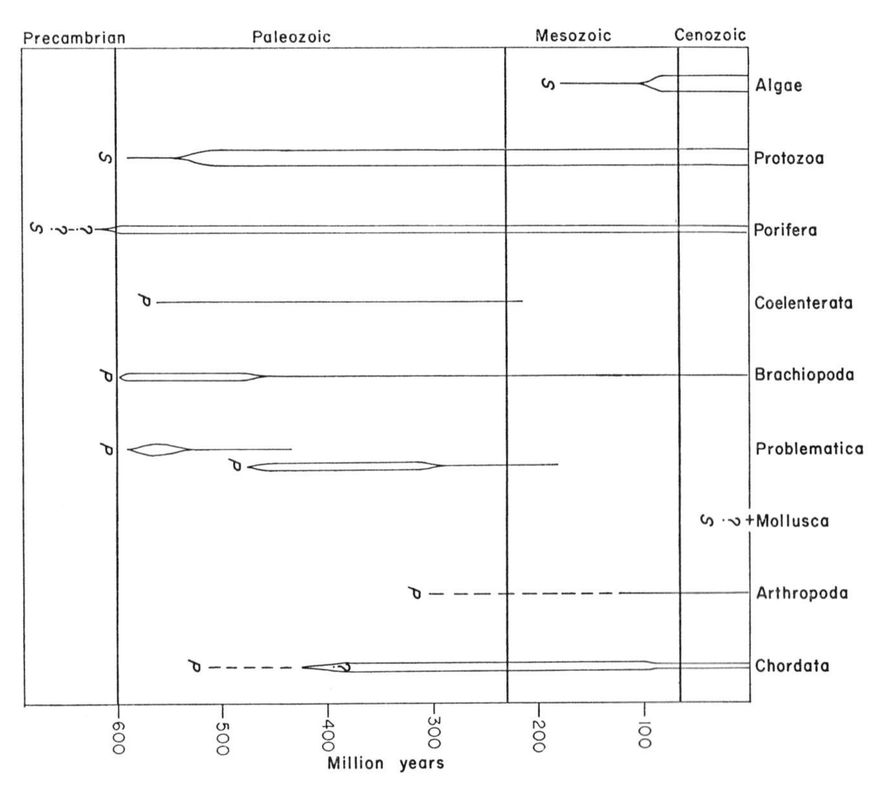

Fig. 10.—Time-stratigraphic distribution of silica (*S*)- and phosphate (*P*)-secreting organisms. Width of bars indicates relative importance of groups.

time. The additions to the silica-secreting organisms in the mid-Mesozoic encompass largely planktonic organisms. Hence the evolutionary history of silica secretion by marine organisms indicates a decided trend toward increased silica extraction from sea water by planktonic over benthonic elements. Beginning with the early Tertiary, shortly after the effective penetration of the euphotic surface waters by the silica-extracting diatoms and flagellates, the siliceous sponges of the shelf bottoms became drastically reduced in class representation, population size, and degree of silicification. The Hexactinellida disappeared from the shelf-sea records and the siliceous Demospongia became minor community constituents and secreted less silica on the average than the pre-Tertiary species. This is reflected in the sedimentary rocks. Cherts derived from sponge spicules, common in the pre-Tertiary marine deposits, are replaced in the Tertiary by diatomites. The displacement of the sedentary siliceous sponges by the prevalently planktonic diatoms on the shelfs also resulted in facies dislocation of shelf deposits rich in biologically derived constituent grains of silica.

We still know very little about the evolution of the Paleozoic, and the end of the Triassic is a period of major regrouping of phosphate-secreting organisms in the oceans. Among the Arthropoda, the Malacostraca make their first appearance.[4] The Chordates show major adaptive radiation among the fish and include the first marine Reptiles. The phosphate-precipitating Coelenterata and Problematica disappear at the end of this time interval. At the beginning of the Jurassic, the phyla with phosphate-secreting organisms that survived had become reduced to the three groups represented today, the Brachiopoda, Arthropoda, and Chordata. Except for the appearance of the phosphate-precipitating tectibranch Gastropoda in the early Tertiary, only minor changes are indicated for the time interval between the beginning of the Jurassic and the present. These changes are the adaptive radiation of phosphate-precipitating Decapoda (Malacostraca), reaching an apparent climax in the Cretaceous, and the addition of marine mammals in the early Tertiary.

These data show that biosynthesis of phosphate minerals as skeletal-strengthening material shifted from an initially wide spectrum of phyla that included organisms of lower structural complexity to a narrower one centered on more advanced structural grades. In a consideration of the modes of life, it is interesting to note that in phyla where phosphate secretion was abundant, most of the organisms belonged to the sessile benthos. The survivors of the early phase of phosphate secretion were almost all parts of the vagile benthos and developed largely into nectonic elements that were joined by other vagile benthonic and nectonic newcomers to the synthesis pool.

Figure 11 shows the time-stratigraphic distribution of the carbonate-secreting organisms. The data indicate early saturation on the phylum level of most major taxa engaged in carbonate precipitation today. Sub-

[4] Possible earlier phosphate-precipitating Arthropoda are indicated by some Cambrian Branchiopoda whose exoskeletons have been reported to consist of this mineral substance (Ulrich and Bassler, 1931). These occurrences need further checking.

tractions and additions are largely limited to class, order, and family levels. Possible significance may be attached to the fact that the earliest carbonate-precipitating marine animals are representatives of the more advanced grades of tissue differentiation among the invertebrates. By contrast, the two lowest grades of organization, the Protozoa and Porifera, seem to have acquired carbonate metabolism considerably later. Unquestionable calcareous Foraminifera make their first appearance in the fossil records in late-Devonian time and the Calcispongia in mid-Devonian time. This favors the thesis held by some that evolution of

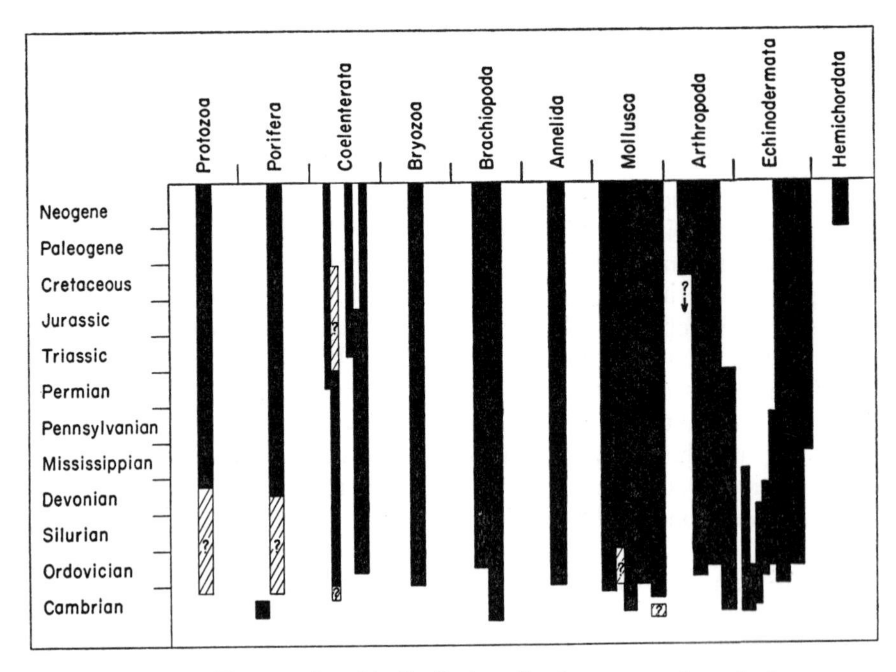

FIG. 11.—Time-stratigraphic distribution of carbonate-secreting animals

marine invertebrates must have reached the stage of complete phylum differentiation prior to the beginning of the Cambrian. The silica- and phosphate-secreting organisms similarly make their first appearance in early Cambrian time, and some of the earliest phosphate-precipitating organisms belong to the more advanced grades of tissue differentiation. This strongly favors the interpretation that invention of diverse modes of mineral metabolism as an evolutionary phenomenon was solely responsible for the explosive expansion of life in the early Cambrian records.

Unquestionably, biogenic carbonate deposits are known from as far back as the late early Cambrian and begin to be common from the late

Cambrian on. We are all familiar with the fact that the major sediment-contributing organisms have changed in the course of geologic time from the early Cambrian on. It is interesting to inquire about particular changes that occurred in the course of evolution of the carbonate-secreting biomass, as it affected the turnover in major biologic contributors of carbonate to the marine sediments, in order to determine whether one can discern a trend in the order in which groups of organisms replaced each other. It is quite clear that no quantitative data exist to investigate this question, and hence one has to fall back on reasonable guesses as to the importance of different groups of organisms that replace each other as major carbonate contributors to the sediments during the course of geologic time. Figure 12 is an attempt to show the time-stratigraphic distribution of those groups of organisms which at one time or another constituted major sediment contributors of carbonate on a world-wide basis. The widths of the bars are highly qualitative estimates of their relative importance as sediment contributors. Taxonomic differentiation is on the order to phylum level, depending on the degree of diversity of the contributors. The arrangement of the taxa from left to right is determined by the time of their greatest importance as major sediment contributors from the Cenozoic to the Paleozoic. The graphic presentation in figure 12 indicates a progression of changes in time of the first- and second-order carbonate contributors as far as their phyletic relations are concerned. In the early Paleozoic, Pelmatozoa and articulate brachiopods constitute first-order contributors, and cephalopods, stromatoporoids, bryozoans, corals, and trilobites, second-order ones. On the average, stromatoporoids may turn out to be only third-order contributors. In the late Paleozoic, principal changes concern the second-order contributors. Cephalopods and corals became reduced to the rank of third-order contributors, and trilobites and stromatoporoids became insignificant sedimentary components. Their place as second-order contributors is taken over by the gastropods, pelecypods, Foraminifera, and possibly by the algae. The Mesozoic, from the Triassic through the early Cretaceous, is marked by a major change in major carbonate contributors. The first-order contributors of the Paleozoic—the Pelmatozoa and articulate brachiopods—became reduced to the rank of minor sediment-forming organisms. Algae, pelecypods, cephalopods, and even corals and Foraminifera form biogenic carbonate deposits, singly or jointly, in very local environment-controlled facies. However, there are no comparable first-order contributors in the designated Mesozoic time interval to take the place of the Pelmatozoa and articulate brachiopods of the Paleozoic, when considered on a world-wide basis. Instead, a considerable variety of second- and third-order carbonate contributors appear to furnish

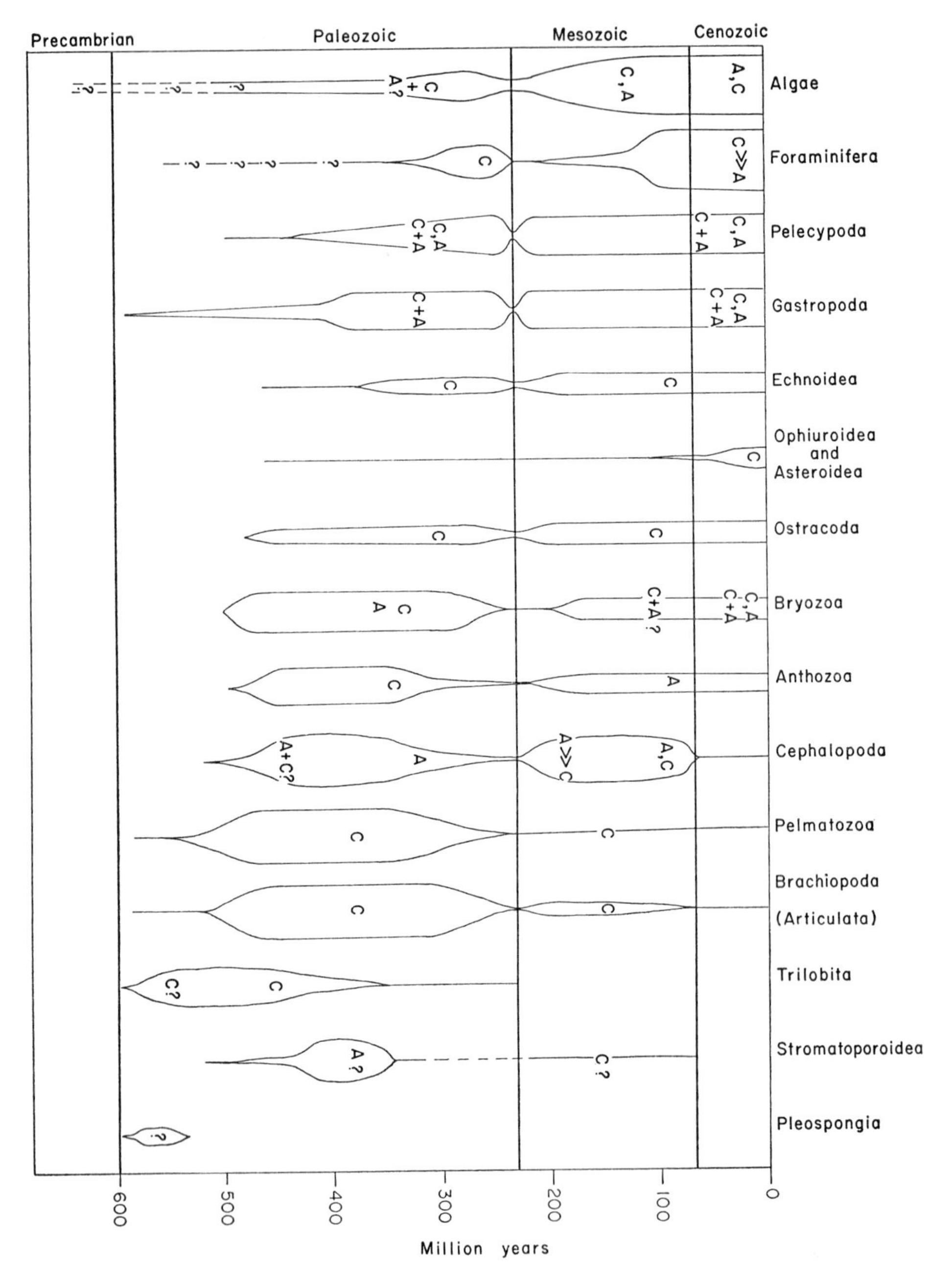

FIG. 12.—Time-stratigraphic distribution of important carbonate-secreting organisms. Width of bars indicates relative importance as contributors to sediment. C = Calcite; A = Aragonite.

skeletal carbonates to the sediments, as shown in figure 12. The late Cretaceous marks the beginning of a third phase in regrouping of major carbonate contributors which seems to have been essentially maintained up to the present. A group of first-order contributors is re-established and represented by the calcareous algae and Foraminifera. The pelecypods and gastropods constitute second-order contributors of carbonate.

The assignment in rank of the algae and Foraminifera as first-order contributors since the late Cretaceous is justified by taking into account the fact that the planktonic members of these two groups of organisms are major carbonate contributors to the pelagic sediments, as documented by oceanic cores as far back as the Eocene and by Cretaceous neritic shelf sediments. The groups considered as first- and second-order contributors of carbonates in pre–late Cretaceous time refer to shelf-sea records, and the organisms in question are largely benthonic elements; only a few were nectonic with unknown oceanic ranges. This emphasizes the great void in our knowledge of the carbonate-secreting organisms and, for that matter, of all mineral-synthesizing biologic agents for the greater parts of the oceanic deposits of pre-Tertiary time. Further, it introduces an element of inequality into the assignment of ranks for the carbonate contributors of pre- and post-Tertiary time. Normalizing the pre- and post–late Cretaceous data by considering biogenic shelf-sea carbonates only, the Foraminifera would still constitute first-order contributors, largely because of the abundance of benthonic species. In the case of the algae, the exclusion of most of the depositional records of the planktonic elements would more likely reduce their contributions on the shelfs to a second-order element.

In examining the changes in major carbonate contributors to the shelf deposits from the early Paleozoic to the present, one finds among the invertebrates a gradual displacement from representatives of the more advanced grades by intermediate and lower grades of structural complexity, compensated in part by increasing importance of the calcareous algae. There is also a trend from prevalently sessile-benthonic, to vagile-benthonic, nectonic, and finally to planktonic elements.

The highly qualitative nature of the data has been stressed, and assignments in rank of second-order, and to some extent first-order, contributors are likely to be revised once quantitative data become available. The primary purpose for presenting this discussion is to stress the decided need for quantitative data on the subject and the intriguing possibilities they offer in defining trends and changes in contributions by the carbonate-secreting organisms to the marine sedimentary rocks and their diagenetic alteration during the course of geologic time.

SKELETAL MINERALOGY IN SEDIMENTARY ROCKS

The skeletally derived minerals and mineralized skeletal remains in the marine sedimentary rocks are commonly altered by diagenetic processes, largely because of contact with fresh water and as the result of surface weathering processes on land. The older the marine sedimentary rocks, the more often diagenetic changes are encountered, but the degree of such alterations is dependent upon the mineralogy and permeability of the rocks. Hence some Paleozoic mineralized remains in some facies may be less altered than Pleistocene remains in others. The diagenetic changes of opaline skeletal remains are in the direction of dehydration of the silica to low-temperature crystobalite and microcrystalline quartz. These remains may become reorganized shortly after accumulation into cherts, which undergo the same sequence of changes. The calcium hydroxyapatites become also diagenetically dehydrated and altered to calcium fluorapatites, as indicated by bones and tooth enamel of vertebrates, for which there is some data on their original mineralogic composition. Studies by X-ray diffraction of the problematic phosphatic remains of conodonts show that they consist also of minerals of the apatite groups (Ellison, 1944; Hass and Lindberg, 1946; Rhodes, 1954). Skeletal carbonates originally composed of aragonite tend to become diagenetically replaced by calcite, and the amounts of original aragonitic constituent grains in fossil carbonates decrease rapidly with increase in geologic age. The oldest known aragonitic skeletal remains are from the Lower Carboniferous (Hallam and O'Hara, 1962). Recent experimental studies have shown that high-magnesian calcites have a higher solubility than aragonites, and hence it would appear that their susceptibility to diagenetic processes should be high and the incidence of their original preservation low compared to low-magnesian calcites and aragonites of increasingly older sedimentary rocks (Chave *et al.*, 1962). Sedimentary carbonates originally composed of high- and low-magnesian calcites and aragonites from skeletal remains tend, especially in older rocks, to be replaced by dolomite.

As indicated earlier, we still know very little about the original mineralogy and chemistry of the siliceous and phosphatic skeletons of recent marine organisms, and hence it is difficult to distinguish between original mineralogy and minor diagenetic alterations in fossil remains composed of similar chemical compounds. Mineralized remains composed of oxides and sulfate minerals have not been reported from the fossil marine sedimentary rocks, indicating another major area of uncertainty that remains to be investigated. There are data on carbonate mineralogy and on some criteria to distinguish original from diagenetic alteration products in fossil skeletal remains. It is of interest to consider (1) methods of differentiating original from diagenetically altered carbonate minerals, (2) criteria to relate diagenetically altered minerals to the original mineral phase, and (3) inferences from these data on the distribution in carbonate minerals in fossil skeletons as they concern changes related to evolution, and of evolution on major sedimentary carbonate contributors.

It is to the credit of Bøggild (1930) that a method has become available to distinguish the original and diagenetically altered mineralogy in skeletal carbonates in certain facies of the sedimentary rocks from the geologic past. The method is based on comparisons of fossil carbonate assemblages in which skeletal aragonites are entirely preserved with assemblages composed of the same species but where part or all of the original aragonite is replaced by secondary calcite or is preserved in the form of casts and molds. As indicated earlier, Bøggild also investigated skeletal microstructure, in particular of the Mollusca, and demonstrated that the crystal fabric was characteristic for the species. By comparing in fossil species the crystal fabric of the original and diagenetically altered crystals, it was shown that the calcitic layers of shells retained their original microstructure, whereas secondary calcite, which replaced aragonitic shells or shell layers, was found to consist of large, blocky, crystallographically randomly oriented crystals, which can be readily distinguished by fabric and orientation from the original aragonite crystals. The obliteration of the skeletal aragonitic microstructures through replacement by secondary calcite was shown to be a general, consistently observable phenomenon. Bøggild found original skeletal aragonites in fossils only as old as Triassic. Utilizing the criteria which allowed him to infer the original carbonate mineralogy in post-Permian fossils, even when original aragonites were replaced by secondary calcite, Bøggild diagnosed the original mineralogy of a number of carbonate fossils from the Paleozoic. Preservation of the microstructure and crystal fabrics in skeletons of the extinct tabulate and rugose corals, trilobites, and certain monoplacophoran orders was interpreted to indicate original calcitic skeletons. Late Paleozoic orthocone nautiloids were shown to have a secondary calcite type of microstructure and on this basis were stated to have consisted originally of aragonite. Stehli (1956) has since found late Carboniferous assemblages that include fossils with skeletons still composed of aragonite and that contained aragonitic shells of orthocone cephalopods and calcitic rugose corals and trilobites. Recently Hallam and O'Hara (1962) have extended the time-stratigraphic range of original aragonite preservation in fossils to the Lower Carboniferous. Shells of goniatites and orthocone nautiloids were found to consist of aragonite and those of a zygopleurid gastropod of calcite and aragonite. The distribution of aragonite and calcite shown by the carbonate skeletons of these late Paleozoic assemblages is in agreement with Bøggild's diagnosis, as based on late Paleozoic occurrences, where the skeletal aragonites had been replaced by calcite. Data on the Mg and Sr contents of late Paleozoic fossils, to be discussed later, similarly are in support of the soundness of Bøggild's approach of utilizing the microstructure and crystal fabric for establishing original carbonate mineralogy. It is quite clear that earlier hesitancy to take advantage of this approach in determining the original carbonate mineralogy in extinct Paleozoic groups is unwarranted. Safe application of Bøggild's method requires impermeable shales containing skeletons or structural units of skeletons, which show in thin sections original microstructure and, under polarized light, original crystal fabric and crystallographic orientations. Hudson (1962) has recently shown that the coarse calcite crystals that replace original aragonite in certain fossil shells may show faint outlines of the original crystal fabric because of inclusions of remnants of their structurally preserved organic matrices. As Hudson points out, the secondary origin of the calcite becomes quite clear under polarized light from the faint fabric traces shown by the orientation and boundaries of the calcite crystals. With these qualifications, it should be possible to utilize with confidence Bøggild's method for a systematic study of the original carbonate mineralogy of carbonate fossils, and in particular of the extinct Paleozoic groups.

There are some data to show that the mineralogy in carbonate skeletons in some groups of organisms has undergone changes in the course of their evolution. In the nautiloids the shells of some Ordovician and Silurian species indicate, on grounds of their microstructure, that they consisted originally of an outer layer of calcite and an inner layer of aragonite (Bøggild, 1930). Either the shells of all post-Devonian nautiloids examined to date consist of aragonite or their microstructures indicate that this mineral species was the original shell carbonate. The same is true for the ammonoids, which descended from the nautiloids and ranged to the end of the Mesozoic.[5] The early nautiloid species were benthonic organisms, commonly with narrow habitat limits, as shown by their facies distributions, and the chambers of their shells were apparently not yet filled with gas to give buoyancy to the animal. In contrast, later nautiloids and ammonoids were nectonic organisms with extended ecologic and hence facies ranges. Therefore it is likely that the changes in shell mineralogy in the nautiloids from calcite plus aragonite to aragonite are surface expressions of biochemical evolutionary changes in the nautiloids and not related to ecologic factors.

[5] The mineralized hard parts secreted by tissues other than those which deposited the shells in nautiloids and ammonoids have been reported to contain calcite. These are the mineralized beaks of some Triassic nautiloids and the aptychi and anaptychi of Mesozoic ammonoids (Bøggild, 1930).

The Monoplacophora present a similar case. Three living species of this group thought to have been extinct since the end of the early Paleozoic have recently been discovered in the deep sea (Lemche, 1957; Clarke and Menzies, 1959; Menzies and Robinson, 1961). The shell of the first-discovered species has been shown to consist entirely of aragonite (Schmidt, 1959), and studies in our own laboratory show that the shells from the two other species are similarly composed of aragonite. Bøggild (1930) indicates that he examined thin sections of the shells of some Silurian species and mentions that the microstructure is preserved, and hence the shells consisted of calcite. Shells from species of the Silurian genera *Tryblidium* and *Pilina* from Gotland examined by the writer (1963) in thin sections indicate that they consisted of an outer calcitic and inner aragonitic layer. The Silurian Monoplacophora from Gotland come from algal-coral-stromatoporoid reefs and shallow-water facies nearby, indicating a warm-water environment and precluding the possibility that their shell mineralogy was temperature determined. Therefore, the differences in shell mineralogy between the Silurian and recent species are best interpreted to indicate evolutionary, biochemically determined changes.

The most extensive documentation to date on changes in carbonate mineralogy by different members of a subclass during the course of geologic time is presented by the Zoantharia in the phylum Coelenterata. Data were presented earlier to show that the skeletons of some late Carboniferous rugose corals consisted of calcite. The Rugosa are limited to the Paleozoic. A systematic study of the microstructure from the calcitic skeletons of all families has been carried out by Wang (1950). The samples ranged in geologic age from mid-Ordovician to Permian and were found to preserve distinct, family-defined microstructures and crystal fabrics. This indicates that all Paleozoic rugose corals secreted calcitic skeletons. By contrast, all but a few Cretaceous species of the scleractinian corals, which succeeded the Rugosa in post-Paleozoic time, had aragonitic skeletons like their recent representatives (Bøggild, 1930). Wells (1954) considers it probable that the scleractinians were derived from soft-bodied anemones and not from the rugose corals. If this is true, the mineralogic differences between the Paleozoic and post-Paleozoic corals are related to physiologic differences of genetically distinct stocks that acquired carbonate metabolism at different times. Hence these mineralogical differences cannot be related to biochemical changes in a single evolving stock, as seems to be the case in the nautiloids and Monoplacophora. The changes in skeletal carbonate mineralogy that occurred during the course of the evolution of the nautiloids and Monoplacophora, and in the replacement of the Paleozoic rugose corals by the post-Paleozoic scleractinians, are similar in that they both result in the substitution of aragonite for calcite in the mineralized skeletons. This trend has been noted earlier for some of the examples cited here (Lowenstam, 1954*c*). There are some indications that this is not a rare phenomenon among carbonate-secreting organisms.

Recent studies of the carbonate mineralogy in calcareous Foraminifera have shown that, aside from calcite, which was known to be the common carbonate phase precipitated by these organisms, aragonite is found also to form their test in some recent and fossil species (Bandy, 1954; Troelsen, 1955; Reichel, 1956; Blackmon and Todd, 1959). Species with aragonitic test are now known to range back to the early Jurassic. In some groups new families were added in which all species precipitate aragonitic tests; in some other pre-existing families with calcite secretion newly evolving species in separate genera began to lay down aragonitic tests. More data on the mineralogy of recent and fossil skeletons are needed to determine whether substitution or evolutionary replacement of aragonite for calcite has been a common phenomenon in the course of evolution of the carbonate-secreting biomass. It is important to consider the implications of this question on possible changes in the amounts of aragonite and calcite that were contributed by the carbonate-secreting organisms to the marine sediments in the geologic past. The nautiloid species, Monoplacophora, and the Foraminifera that are known to have undergone evolutionary changes in carbonate mineralogy constitute minor contributors to the sedimentary carbonates, whereas the corals were second- and later third-order contributors of carbonate to the sediments. From these limited data, one gains the impression that the effect of changes in carbonate mineralogy brought about by evolutionary changes on the original aragonite-calcite ratios of sedimentary carbonates in the oceans through geologic time has been minor when the total volume of biogenic carbonates is considered.

It has been shown earlier that different phyla were major biologic contributors to carbonate sediments at different times, and one may ask in what way these changes affected the aragonite-calcite ratios of the accumulating sedimentary carbonates. Figure 12 shows the carbonate mineralogy of the major sedimentary contributors as it is known at present. These data suggest that during the early Paleozoic the aragonite-calcite ratios were low. During the late Paleozoic they became higher, and still higher in the Mesozoic through the mid-Cretaceous. With the rapid expansion of calcitic algae and Foraminifera, and in particular of their planktonic representatives, as major carbonate contributors there is a reversal in aragonite-calcite ratios toward lower ratios, starting with the late Cretaceous time and continuing to the present. Considering shelf carbonates only, the resultant lowering in aragonite-calcite ratios is noticeable but far less pronounced than when the pelagic carbonate deposits are included. The reversal in trend as it affected the aragonite-calcite ratios in the shelf carbonates did not reach proportions comparable to the low aragonite-calcite ratios for the early Paleozoic. The pelagic carbonates today are derived almost entirely from planktonic algae and Foraminifera and hence are for all practical purposes composed of calcite. The same is indicated for the pelagic carbonates of the Tertiary as determined from studies of cores from the deep ocean floor (Arrhenius, 1952). We have no data on the carbonate-secreting organisms of the pelagic realm prior to the Tertiary, and hence there is a major problem as to the aragonite-calcite ratios of pelagic carbonates compared to those indicated for the shelves. This again focuses on the major gap in our knowledge of the pelagic history of biogenic mineral precipitates. On the basis of our present knowledge, it appears that the changes in aragonite-calcite ratios of shelf carbonates through geologic time were more strongly controlled by shifts in major carbonate contributors than by evolutionary changes in aragonite-calcite precipitation within lineages. Clearly, the information on the subject is incomplete and requires more data to evaluate the controlling factors for the changes in original aragonite-calcite ratios in the carbonate sediments through geologic time.

Sr, Mg, AND O^{18} CONTENTS OF FOSSIL CARBONATE SKELETONS

It was shown that the Sr, Mg, and O^{18} contents of recent skeletal carbonates are affected by temperature and the chemistry of the water in which the carbonate was laid down. It was shown further that in the case of articulate brachiopods it is possible to differentiate between the effects of the two ecologic factors in the Sr, Mg, and O^{18} contents of their shell carbonate. In the application to fossil biogenic carbonates we are interested in knowing whether the chemistry of the oceans has undergone changes with time. There is the further question as to whether biochemical evolution has affected the uptake of Sr and Mg in skeletal carbonates. Skeletal carbonates are subject to diagenetic alterations that may extend only to the chemistry of the carbonate crystals (Lowenstam, 1961). Hence it is of utmost importance to differentiate between original and diagenetically altered Sr, Mg, and O^{18} contents of fossil skeletal carbonates. Data on the original Sr and Mg contents in fossil skeletal carbonates are needed to detect possible changes in the course of geologic time, which in turn may have a bearing on diagenetic alterations of biogenic carbonate deposits.

The question whether it is possible to distinguish original from diagenetically altered Sr and O^{18} contents of fossil skeletal carbonates is considered first. Figure 13 shows the relations of the $SrCO_3$ contents to the O^{18}/O^{16} ratios in samples of shells from articulate brachiopods

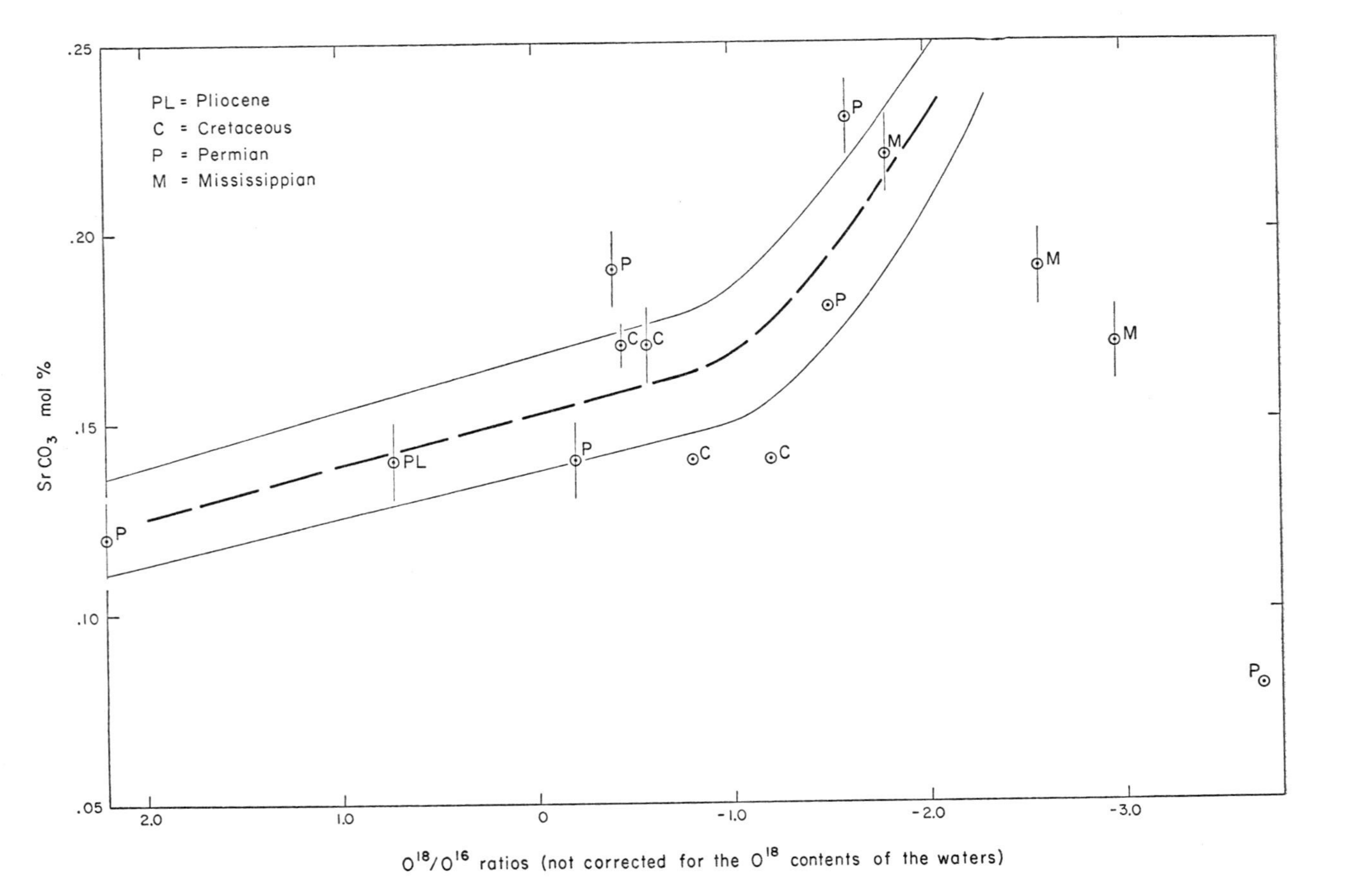

FIG. 13.—Plot showing $SrCO_3$ content against O^{18}/O^{16} ratios for fossil brachiopods

ranging in geologic age from the late Pliocene to the late Mississippian (Lowenstam, 1961). The points for the Pliocene and some for the Cretaceous, Permian, and Mississippian samples are shown to be close to those for recent species from waters within ± 5 per cent of mean ocean salinity. The close agreement in relations between the $SrCO_3$ contents and the O^{18}/O^{16} ratios of these fossil samples to those of the recent ones indicates that the original $SrCO_3$ contents and O^{18}/O^{16} ratios in these fossil samples are preserved. They indicate further that the Sr/Ca ratios and the O^{18}/O^{16} ratios in the oceans remained the same for the last 3×10^8 years. The relationships of the $SrCO_3$ contents to the O^{18}/O^{16} ratios shown by the three Mississippian samples strengthen this interpretation. The samples are shells of individuals of the same species taken from the same deposit at the same locality. They represent a series of unweathered, slightly weathered, and extensively weathered shells. Compared to the unweathered sample, the relationships of the O^{18}/O^{16} ratios to the $SrCO_3$ contents from slightly and from more extensively weathered shells indicate progressive stages of increasing diagenetic alteration by waters with lower O^{18} and Sr contents than those from mean ocean water. It would seem likely that if the $SrCO_3$ and O^{18} contents of the Pliocene to Mississippian samples, which show similar relations to those for recent species, were diagenetic alteration products, the experimental points for the fossil samples should be randomly distributed relative to those for the recent samples. Changes in Sr/Ca ratios and O^{18} contents in sea water were shown for recent samples to be reflected in the Sr/Ca ratios and O^{18}/O^{16} ratios of the brachiopod shells. Hence, if the Sr/Ca ratios and O^{18}/O^{16} ratios of the oceans had undergone changes in the last 3×10^8 years, one would expect systematic changes in the relationships of the experimental points for the Pliocene to Mississippian samples relative to those for the recent ones. Since neither is the case, it would appear that the original Sr/Ca ratios and O^{18}/O^{16} ratios in the fossil shells are preserved and that the O^{18} contents and Sr/Ca ratios in sea water had not changed within the time span covered by the samples. These data also indicate that the degree of discrimination against Sr relative to Ca as shown in recent shells of articulate brachiopods relative to sea water has remained the same for the last 3×10^8 years.

The relationship of the $MgCO_3$ contents to the O^{18}/O^{16} ratios was also examined for the shell carbonate of the same brachiopod samples (Lowenstam, 1961). The data showed that the Mg contents of the originally high-magnesian calcites were diagenetically reduced to low values, and only in a Pliocene and a Permian sample, where the original magnesian contents were low, were the original concentrations preserved. While these data suggest that the Mg/Ca ratios in the oceans, at least since early Permian time, were similar to those in sea water today, the documentation is less satisfactory than one would like to have for securely establishing this point.

The Paleozoic carbonates from biologic sources are composed of skeletal remains of organisms which either became extinct at the end of this era or are genetically widely removed from recent species in the same classes as phyla. Although we do not at this time have studies of their Sr and Mg contents comparable to those we have for the articulate brachiopods, it is important to explore the Sr and Mg contents preserved in the skeletons of other major contributors of carbonate to the Paleozoic sediments, since one cannot assume that their Sr and Mg contents were similar to those of the carbonate from recent species within the same phyla. To assure a minimum of diagenetic alterations, the samples selected for such an investigation would have to show preservation of the original microstructure and should be derived from sediments of low permeability, such as shales, where fluid migration has been minimal since they were deposited. The preservation of aragonitic skeletal remains would indicate that these conditions had been met.

Calcitic skeletal remains from Tertiary, Mesozoic, and late Paleozoic fossils (from deposits with the qualifications indicated above) were selected for Sr and Mg determinations. Figure 14 shows the Sr/Ca ratios of a variety of aragonitic fossil types in relation to those for recent calcitic samples from the same phyla. Samples of rugose corals and cryptostome bryozoa extinct since the end of the Paleozoic are included. The Sr/Ca ratios of some of the samples from the Rugosa, Pennatulacea, Cyclostomata, Crinoidea, and Echinoidea are within the range of values for recent species, whereas others have lower values indicating loss of Sr due to diagenesis.

Table 1 shows the maximum values of Mg in biologic calcites in solid solution, as determined from X-ray spacing, calculated as mole per cent of $MgCO_3$. The Mg determinations are for the same samples as those for the Sr/Ca ratios shown above and include also data on the extinct fusulinids among the Foraminifera and trilobites among the Arthropoda. The ranges of $MgCO_3$ contents for recent representatives of the same or related higher taxa are presented for comparison. The $MgCO_3$ contents range from 2 to 10–12 mole per cent. The highest values are for samples from the late Paleozoic. It has been shown that skeletal calcites from pre-Tertiary carbonates and carbonate-rich sedimentary rocks are generally practically free of Mg, regardless of whether they were originally high-magnesian calcites or not (Chave, 1954; Goldsmith and Graf, 1955). Echinoid spines of Pennsylvanian age reported to contain 3 mole per cent $MgCO_3$ were collected by the writer from impermeable, carbonate-poor shale (Goldsmith, 1960). The maximum

PLATE I

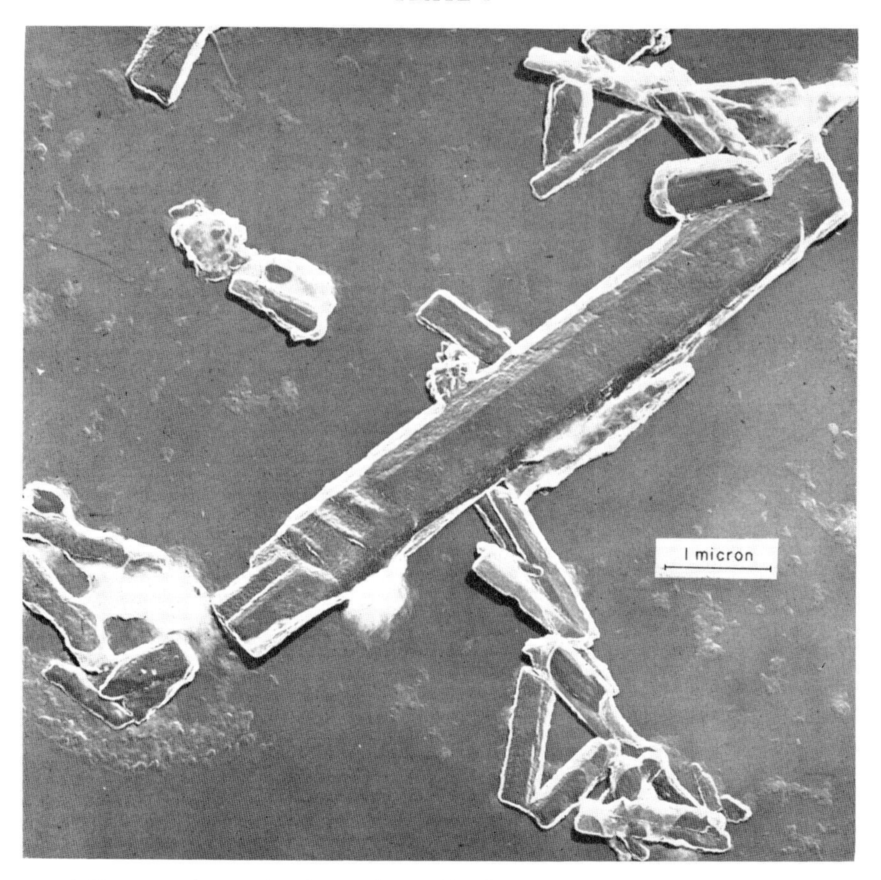

1, Electron micrograph of aragonite needles from the algal genus *Rhipocephalus*

PLATE II

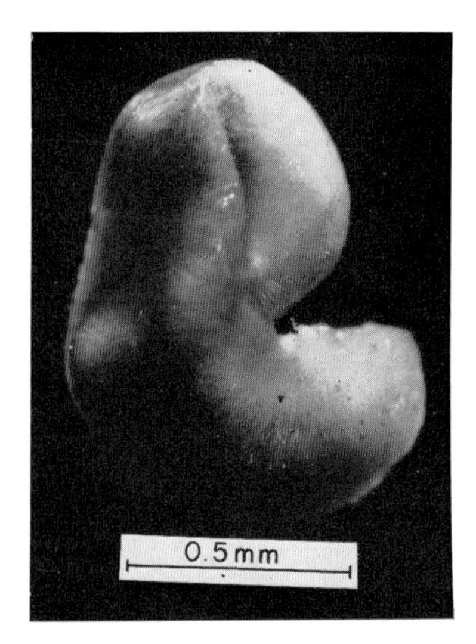

2, Photomicrograph of unidentified Aplacophora with aragonite dermal spicules. USARP Research Vessel, "U.S.N.S. Eltanin," Cruise 3, Peru-Chile Trench; depth, 6,233 meters.

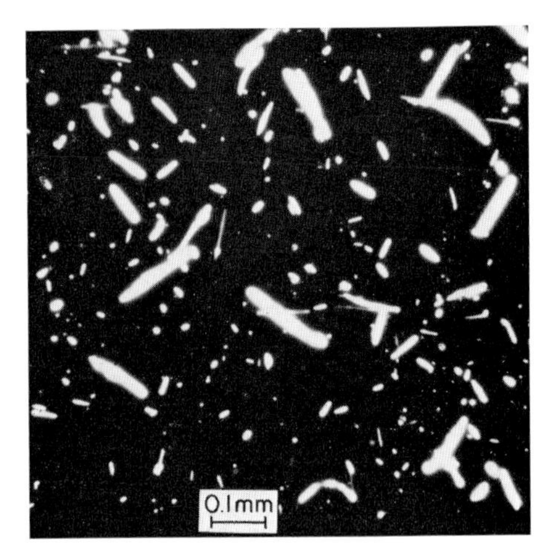

3, Photomicrograph of aragonite spicules from specimen shown in *2*, polarized light.

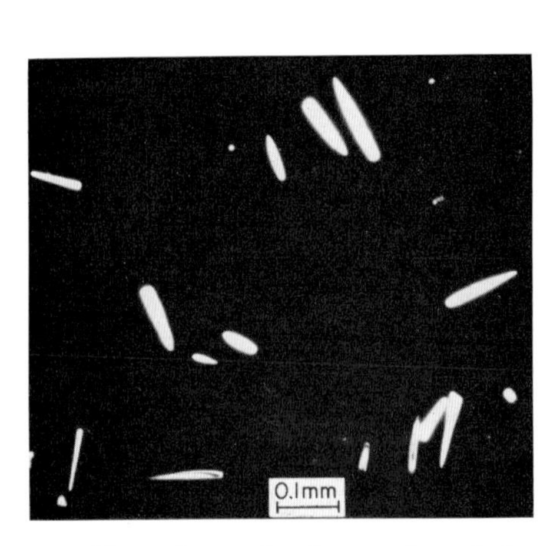

4, Photomicrograph of aragonite spicules from another unidentified Aplacophora, from same cruise as specimen shown in *2;* depth, approximately 2,700 meters, polarized light.

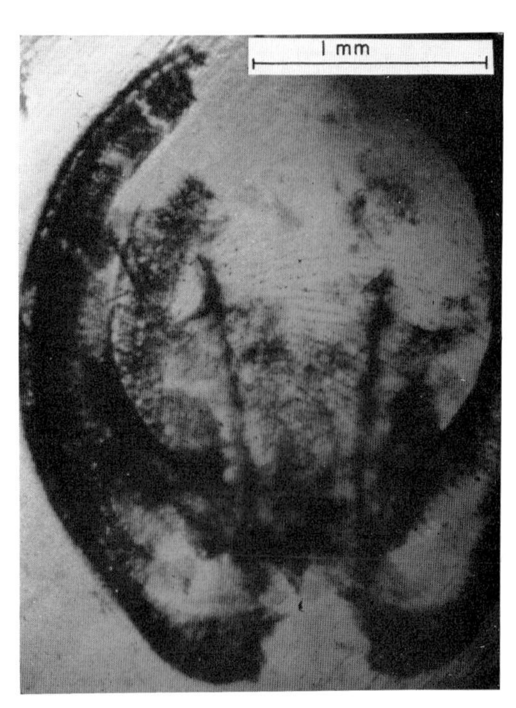

5, Photomicrograph of calcitic spicules on the brachia from the lophophore of *Terebratulina unguicula*, off Juneau, Alaska; depth, 14 meters.

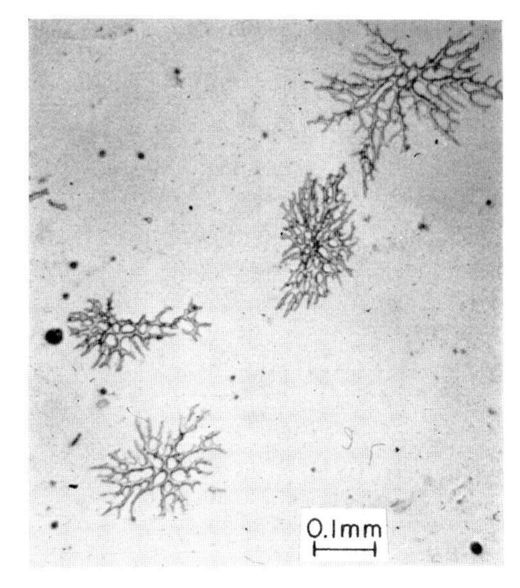

6, Photomicrograph of isolated spicules from specimen shown in *5*.

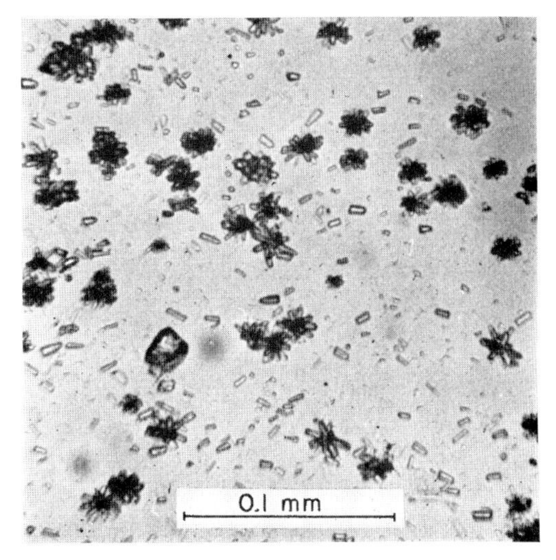

7, Photomicrograph of aragonitic tunicate spicules, family Didemnidae, Palau, Caroline Islands.

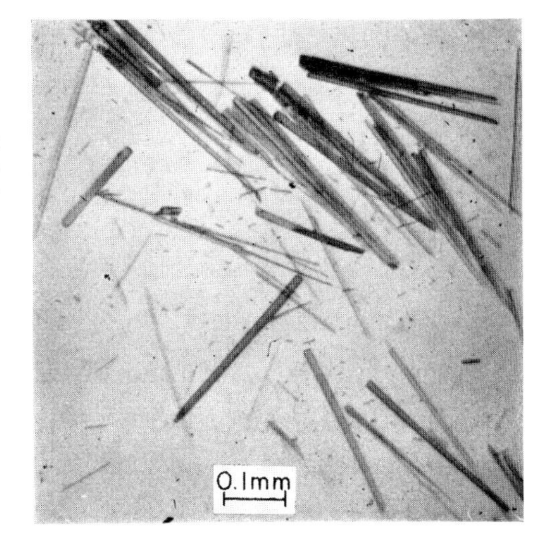

8, Photomicrograph of aragonitic tuft spicules of the chiton *Acanthochiton communis*, Ville Franche, France, Mediterranean.

PLATE IV

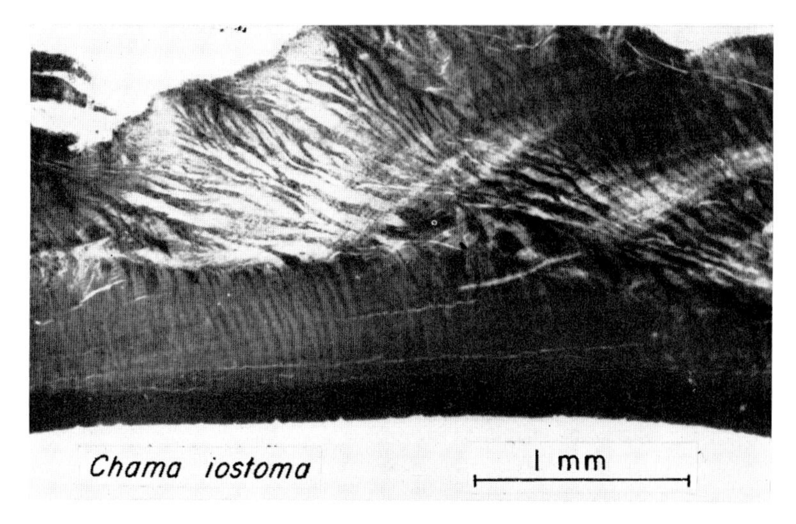

9, Photomicrograph showing microstructure in transverse thin section through shell of the warm-water pelecypod *Chama iostoma*, with outer and inner aragonitic shell layers.

10, Same as *9*, showing cold-water species *Chama pellucida*, with outer calcitic and inner aragonitic shell layers.

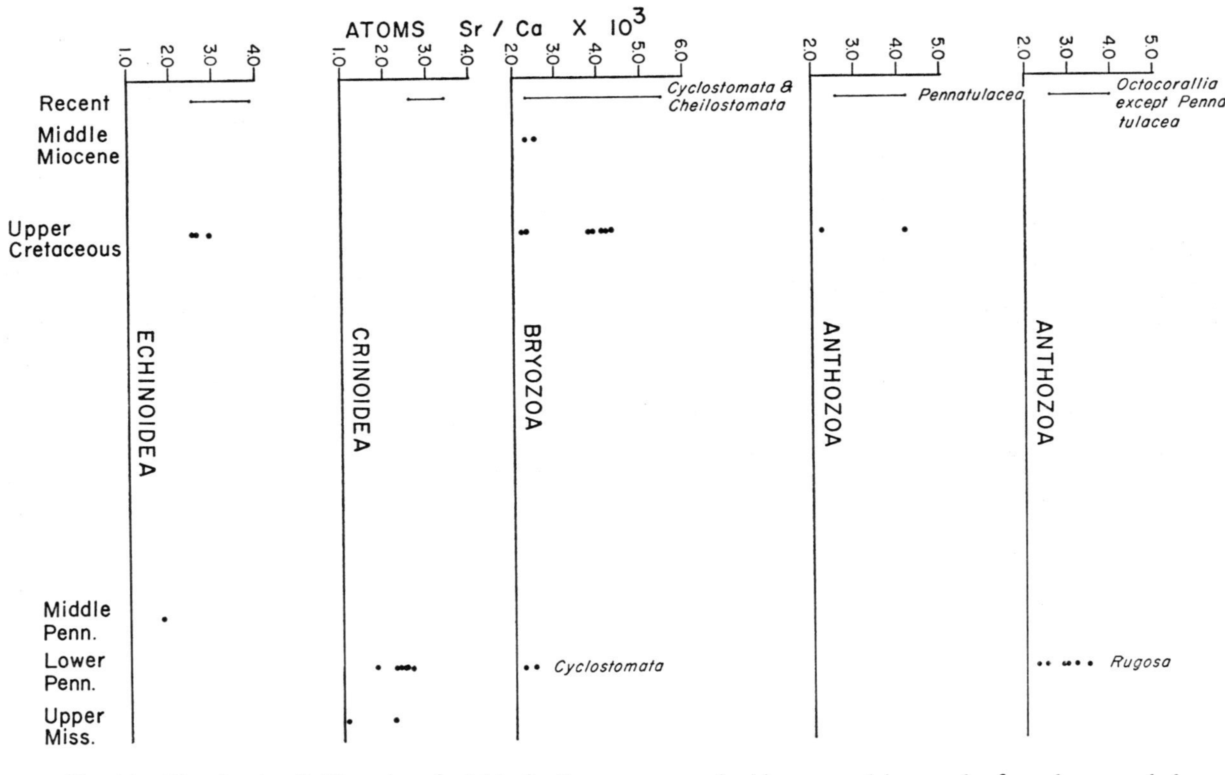

FIG. 14.—Plot showing Sr/Ca ratios of calcitic fossil types compared with recent calcite samples from the same phyla

$MgCO_3$ contents of the recent samples shown in table 1 are far above the values reported for biogenic calcites from pre-Tertiary deposits. The $MgCO_3$ contents for most of the fossil samples shown in table 1 are within the range of values reported for recent species within the orders, classes, and phyla to which the fossils taxonomically belong (Chave, 1954; Goldsmith and Graf, 1955; Blackmon and Todd, 1959). The

TABLE 1

COMPARISON OF $MgCO_3$ CONTENTS OF RECENT AND FOSSIL CALCITE SAMPLES FROM VARIOUS PHYLA

Recent Group	$MgCO_3$ (Mole Per Cent)	Fossil Group	Age	Maximum Values $MgCO_3$ (Mole Per Cent)
Protozoa:				
Foraminifera	0–20	Fusulinidae	M. Pennsylvanian	3
Coelenterata:				
Octocorallia	6–14	Pennatulacea	Maestrichtian U. Cretaceous	6
		Rugosa	L. Pennsylvanian	4
Bryozoa	5–12*	Cheilostomata	M. Miocene	7
		and Cyclostomata	Maestrichtian U. Cretaceous	6
		Cryptostomata	L. Pennsylvanian	6
Arthropoda	2–14*	Trilobita	L. Pennsylvanian	5
Ostracoda	2–11			
Decapoda	6–14			
Echinodermata:				
Crinoidea		Inadunata	L. Pennsylvanian	10–12
Articulata	9–16	?	M. Pennsylvanian	3
		Inadunata	U. Mississippian	4
Echinoidea	5–16	Irregularia	Maestrichtian U. Cretaceous	8
		Regularia:	M. Pennsylvanian	
		test plates		2
		spines		3

* Data incomplete.

$MgCO_3$ contents of the fossil samples clearly indicate that their skeletons were originally composed of calcite with appreciable amounts of Mg. However, since the Mg contents of high-magnesian calcites tend to be reduced during diagenesis, there is no assurance that the concentrations determined for the fossil samples represent the original concentrations.

The fusulinids, rugose corals, cryptostome bryozoans, and trilobites became extinct at the end of the Paleozoic. The preservation of their skeletal microstructure and their burial association with skeletal carbonates composed of aragonite strongly favor the view that they were composed originally of calcite, as stated earlier for some of these forms by Stehli (1956). Showing here that their $MgCO_3$ contents were appreciably in excess of those found in secondary calcites of unmetamorphosed carbonate deposits provides further independent evidence of their original mineralic composition. In the case of the rugose corals the data on their Sr/Ca ratios presented here and those previously reported by Odum (1957) strengthen the argument that they had calcitic skeletons. Similarly, the relation of the Sr/Ca ratios to the $MgCO_3$ contents of carbonate skeletons from the extinct cryptostome bryozoans can be explained only if the carbonate consisted originally of calcite. It is interesting to note that some of the crinoidal calyx plates from the Pennsylvanian contain on the average 10 mole per cent $MgCO_3$ in solid solution and have local patches with up to 12 mole per cent $MgCO_3$. Traverses of the crystals with an electron microprobe should determine whether the highest locally preserved $MgCO_3$ contents fall within or lie above the range of values found in recent crinoid plates. The same applies to the calcites of other Paleozoic fossils where X-ray diffraction indicates a range in $MgCO_3$ contents. The investigation of this question is important in order to determine whether the Mg uptake in the carbonate of high-magnesian calcite-precipitating organisms has undergone evolutionary changes. It is also important from the sedimentary point of view to assess the contribution of the exsolution of Mg from biologically precipitated high-magnesian calcites through exsolution to the formation of dolomites as diagenetic alteration products of sedimentary carbonates.

It has been pointed out earlier that calcite-precipitating organisms were the major contributors of carbonates to the shelf-sea deposits during the early Paleozoic (fig. 12). These major sediment contributors included crinoids, trilobites, rugose corals, articulate brachiopods, and bryozoans. The Mg determinations for samples of some of their late Paleozoic representations indicated that these organisms deposited medium- to high-magnesian calcites and that the actual original Mg contents may have been higher than those determined. The proportion of medium- and high-magnesian calcites in the sedimentary carbonates decreased subsequently in pre–late Cretaceous time due to extinction of some of their major contributors at the end of the Paleozoic and as a result of increase in aragonite-precipitating groups of organisms, in particular of the Mollusca. Skeletal aragonites are low in Mg contents, and molluscan skeletal calcites are similarly low in Mg. The increased importance of the carbonate-secreting Foraminifera in post-middle Cretaceous time added a major contributor of high-magnesian calcites to the benthonic fraction, and a low-magnesian calcite contributor to the planktonic fraction of the sedimentary carbonates. The increased importance of planktonic calcareous algae added still another major contributor of low-magnesian calcite to the carbonates of post-middle Cretaceous time.

Hence the amounts of high-magnesian calcites contributed to the younger sedimentary carbonates were not as great as those in the early Paleozoic. The addition of two major biologic sources of low-magnesian calcites to the one already existing—the Mollusca—would seem to suggest a progression from high- toward low-magnesian calcites in the marine sedimentary carbonates during the course of geologic time. The increase of the planktonic Foraminifera and algae has been recognized for some time (Twenhofel, 1930; Kuenen, 1950) and was thought to account for the lesser amounts of dolomite found in post-Paleozoic carbonate rocks. As has been pointed out repeatedly here, one must distinguish between the history of shelf and pelagic carbonate deposits, since there are no data on the carbonate-contributing organisms of pelagic sediments of pre-Tertiary age. It is true that the fraction of biogenic high-magnesian calcites in the shelf carbonates has decreased, for reasons pointed out, since the late Paleozoic. To relate causally the post-Paleozoic decrease of dolomite to a decrease in biologic sources of high-magnesian calcites implies that most if not all of the excess in Mg required to produce these dolomites had to come from exsolution of Mg from skeletal carbonates. The calcites of crinoidal limestones from the Paleozoic have been commonly diagenetically altered to low-magnesian calcites, and hence there has been mass transfer of Mg from these originally high-magnesian calcites to other sources. As indicated earlier, more data are needed to determine the original Mg contents of the major biologic contributors of medium- to high-magnesian calcites to the carbonate deposits of the Paleozoic, and hence it is too early to propose a materials balance showing how much of the Mg in dolomites is derived from biologic and how much possibly from other sources, as, for example, connate brines. In the case of the pelagic carbonates, we know only that, beginning with the Tertiary, low-magnesian calcites became the principle contributors of marine organisms of carbonates to the sediment in deep water. Above the depth at which aragonite is dissolved, smaller amounts of aragonite were added by pteropod skeletons. Data from the future Moho drillings will be required to determine what carbonate phases were contributed by organisms to four-fifths of the ocean floor during the Paleozoic and the pre–late Cretaceous Mesozoic.

Returning to consideration of the shelf carbonates, it has been shown that the Mg concentrations in calcitic skeletons increase with elevation in environmental temperatures. For geologic periods with a more even climate than that of today the amounts and absolute concentrations of Mg of most biologic calcites in the sedimentary carbonates for the shelfs as a whole should have been greater than those of today. Paleotemperatures can now be determined with greater confidence by the methods indicated earlier. It is therefore possible to investigate the history of paleoclimates and determine the changes in areal extent of the tropical and polar zones. Such studies should aid in tracing the changes in amounts and in concentrations of Mg of high-magnesian calcites of the shelf-sea carbonates in the geologic past.

There is a decided need for data on the chemistry of biologically precipitated silica and of other chemical compounds that contribute parts of the sedimentary rocks of the geologic past.

ORGANIC MATRICES OF FOSSIL SKELETAL CARBONATES

There are some data on the physical and chemical properties of the organic matrices from originally aragonitic carbonate skeletons of fossil molluscs. Decalcification of Mesozoic ammonoid and Cenozoic to Ordovician nautiloid shells has yielded remnants and, in some cases, major portions of the organic matrices that are still physically intact or only slightly altered (Grégoire, 1958, 1959, 1962). Electron microscope studies of the decalcified organic matrices from recent and Pennsylvanian nautiloid shells have shown that they consist of a reticulate network that is quite similar to the recent and the fossil forms (Grégoire, 1959). Abelson (1954, 1959) determined the principal amino acid contents of fossil molluscan shells as old as Miocene. A qualitative study of the amino acid composition from the decalcified organic matrix of a Pennsylvanian nautiloid shell (*Pseudoorthoceras knoxense*), which consists of the original aragonite, has been carried out by Dr. E. Hare. Lysine, histidine, arginine, aspartic acid, tyrocene, serine glutamic acid, glycine, and alanine were found to be present. The ratios of amino acids appear to be similar to those of organic matrices associated with recent calcified molluscan shell material—that is, a high glycine content. The same amino acids were found in the organic matrices of the cameral deposits of this species, and hence the carbonate of the cameral deposits is clearly indicated to be an original carbonate precipitate of these nautiloids.

These data are encouraging, since they indicate that in fossils with original shell mineralogy the organic matrices may be preserved physically intact, and there is the suggestion that, under exceptional preservational conditions involving early sealing of the sediment pores, the amino acids of the insoluble organic matrix material may be mostly preserved. It will remain for future studies to determine whether the physical state and chemical composition of organic matrices from unusually well-preserved carbonate skeletons from different types of fossils are sufficiently close to the original ones to contribute toward defining genetic relationships and elucidating aspects of biochemical evolution. The answers to these questions are dependent, as in the case of the carbonate mineralogy and chemistry, on more data from living organisms from different orders, classes, and phyla. Lack of characterization of all potential source materials and the uncertainty of organic and inorganic processes that cause degradation of the organic fractions as we find them in the fossil sedimentary rocks seem to indicate that it is not yet possible to relate the organic compounds in the sedimentary rocks to their specific biologic sources.

Discussion

The foregoing presentation has touched only on certain aspects that concern the biologic fraction of marine sediments and sedimentary rocks. It was shown that our present knowledge of the kinds of minerals that are precipitated by recent marine organisms is still incomplete, and progress must come from studies of living organisms from the marine biomass rather than from their sedimentary remains. Present lack of information on all existing kinds of minerals precipitated by marine organisms and of their physical and chemical properties introduces in many cases an element of uncertainty into the distinction between biologic and inorganic origins of sedimentary particles. Considerable progress has been made in relating the physical and chemical properties of biogenic carbonates to the biologic and ecologic factors that control their precipitation. There are also some data on the physical and chemical properties of the protein matrices of mineralized carbonate skeletons. Progress in this area should give insight into the mechanisms of skeletal-mineral precipitation and provide data on the biochemistry and ecology of the organisms. It is becoming quite clear that the physical and chemical properties of the organic matrices and of the mineral precipitates must be considered jointly in order to differentiate between the effects from different sources and to evaluate the effect of burial of their preservable fractions. Growth experiments under controlled conditions in the laboratory should aid in distinguishing between different effects and in determining quantitatively the relationships of physical and chemical properties of mineralized skeletons to biologic and ecologic controls. The electron microscope and the electron microprobe are indicated as tools to answer certain of the problems outlined earlier.

Our present knowledge of the physical and chemical properties of biologic mineral precipitates and of their relations of their organic matrices is essentially limited to carbonates. Hence exploratory investigation of biogenic carbonates is more advanced than other aspects of the biogenic fraction in the fossil sedimentary rocks. Considerable progress has been made in distinguishing between original and diagenetically altered carbonates of biologic origin and in relating chemical properties of unaltered skeletal remains to biologic and ecologic factors. The results obtained to date are encouraging, but with better data and new techniques it should be possible to carry the study of the carbonate-secreting biomass and the history of the chemistry of the oceans at least as far back as the early Paleozoic. By analogy it should also be possible to obtain similar data from other kinds of minerals precipitated by marine organisms.

The most outstanding problem that is indicated in the biologic area from the foregoing synthesis of the data is our total lack of information on the history of the mineral-precipitating organisms of the pelagic realm of the oceans in pre-Tertiary time and for the shelf seas in pre-Cambrian time. It is similarly clear that because of our limited knowledge of the composition of the biologic remains and their interaction, as well as their effects on the inorganic fraction of the sediments, the evolution of depositional and diagenetic changes of marine sedimentary rocks as affected by marine organisms is still in its infancy.

ACKNOWLEDGMENT

Dr. H. Lemche called the writer's attention to the gizzard plates in the tectibranch gastropod *Scaphander lignarius* and furnished specimens for their mineralogic investigation. Dr. R. J. Menzies provided the shells of *Neopilina* (*N.*) *veleronis* for mineralogic determinations. The new mineralogic and trace element data are based on studies supported by grants from the National Science Foundation, Nos. G-20187 and GP-321.

REFERENCES CITED

ABELSON, P. H., 1954, Annual Report of the Director of the Geophysical Laboratory, 1953–1954: Carnegie Inst. Wash. Year Book 53, p. 97–101.

——— 1959, Geochemistry of organic substances, *in* Researches in Geochemistry, p. 79–103: New York, John Wiley & Sons.

ARRHENIUS, G., 1952, Sediment cores from the East Pacific, *in* Rpt. Swedish Deep-Sea Exped., 1947–1948, Götesborgs Kungl. Vetenskaps–och Vitterhets-Samhälle., v. 5, 227 p.

BANDY, O. L., 1954, Aragonite tests among the Foraminifera: Jour. Paleont., v. 24, p. 60–61.

BLACKMON, P. D., and TODD, R., 1959, Mineralogy of some Foraminifera as related to their classification and ecology: Jour. Paleont., v. 33, p. 1–15.

BLOCHMANN, P., 1906, Neue Brachiopoden der Valdivia- und Gauss-Expedition: Zool. Anz., v. 30, p. 690–702.

——— 1908, Zur Systematik und geographischen Verbreitung der Brachiopoden: Zeitschr. wissensch. Zool., v. 90, p. 596–644.

——— 1912, Die Brachiopoden der schwedischen Südpolar-Expedition: Wissensch. Ergebn. schwedisch Südpolar Exped., 1901–1903, v. 6, Lief. 7, p. 1–12.

BØGGILD, O. B., 1930, The shell structure of the mollusks: K. Danske Vidensk. Selsk. Skrifter, Naturvidensk. og Math. Afdel., 9th ser., v. 2, p. 233–326.

BRAMLETTE, M. N., 1926, Some marine bottom samples from Pago Pago Harbor, Samoa: Carnegie Inst. Washington Pub. 344, v. 23, p. 1–35.

——— 1958, Significance of Coccolithophorids in calcium-carbonate deposition: Geol. Soc. America Bull., v. 69, p. 121–126.

CHAVE, K. E., 1952, A solid solution between calcite and dolomite: Jour. Geology, v. 60, p. 190–192.

——— 1954, Aspects of the biogeochemistry of magnesium: 1. Calcareous marine organisms: *ibid.*, v. 62, p. 266–283.

——— 1954, Aspects of the biogeochemistry of magnesium: 2. Calcareous sediments and rocks: *ibid.*, p. 587–599.

——— 1960, Carbonate skeletons to limestones: Problems: New York Acad. Sci., ser. 2, v. 23, p. 14–24.

———, DEFFEYES, K. S., WEYL, P. K., GARRELS, R. M., THOMPSON, M. E., 1962, Observations on the solubility of skeletal carbonates in aqeuous solutions: Science, v. 137, p. 33–34.

CLARKE, A. H., and MENZIES, R. J., 1959, *Neopilina* (*Vema*) *ewingi*, a second living species of the Paleozoic class Monoplacophora: Science, v. 129, p. 1026–1027.

CLARKE, F. W., and WHEELER, W. C., 1922, The inorganic constituents of marine invertebrates: U.S. Geol. Survey Prof. Paper 124, 62 p.

CLOUD, P. E., 1962, Environment of calcium carbonate deposition west of Andros Island, Bahamas: U.S. Geol. Survey Prof. Paper 350, 138 p.

DODD, J. R., 1963, Paleoecological implications of shell mineralogy in two pelecypod species: Jour. Geology, v. 71, p. 1–11.

ELLISON, S. P., 1944, The composition of the conodonts: Jour. Paleont., v. 18, p. 133–140.

EPSTEIN, S., and MAYEDA, T. R., 1953, Variations in O^{18} contents of waters from natural sources: Geochim. & Cosmochim. Acta, v. 4, no. 5, p. 213–224.

——— and LOWENSTAM, H. A., 1953, Temperature–shell-growth relations of recent and interglacial Pleistocene shoal-water biota from Bermuda. Jour. Geology, v. 61, p. 424–438.

GEE, H., MOBERG, E. G., and REVELLE, R., 1932, Calcium equilibrium in seawater: Scripps Inst. Oceanogr., Univ. Calif. Bull., tech. ser., v. 3, p. 145–200.

GINSBURG, R. N., 1953, Carbonate sediments: API Proj. 51, Rpt. 8, SIO Ref., p. 26–53.

——— 1956, Environmental relationship of grain size and constituent particles in some south Florida carbonate sediments: Am. Assoc. Petroleum Geologists Bull., v. 40, p. 2384–2427.

GOLDSMITH, J. R., 1960, Exsolution of dolomite from calcite: Jour. Geology, v. 68, p. 103–109.

———, GRAF, D. L., and JOENSUU, O. I., 1955, The occurrence of magnesian calcites in nature: Geochim. & Cosmochim. Acta, v. 7, p. 212–230.

GRAF, D. L., 1960, Geochemistry of carbonate sediments and sedimentary rocks: Part 4-B Bibliography, Illinois State Geol. Surv., Circ. 309, 55 p.

GREENFIELD, L. J., 1963, Aragonite formation by marine bacteria (abs.): Am. Assoc. Petroleum Geologists Bull., v. 47, p. 358.

GRÉGOIRE, CH., 1958, Topography of the organic components in mother-of-pearl: Jour. Biophys. & Biochem. Cytol., v. 3, p. 797–808.

——— 1959, Conchiolin remnants in mother-of-pearl from fossil cephalopoda: Nature, v. 184, p. 1157–1158.

———, DUCHATEAU, GH., and FLORKIN, M., 1955, La trame protidique des naces et des perles: Ann. Inst. Oceanogr., v. 31, p. 1–36.

HALLAM, A., and O'HARA, M. J., 1962, Aragonitic fossils in the Lower Carboniferous of Scotland: Nature, v. 195, p. 273–274.

HARE, P. E., 1961, Variations in the composition of the organic matrix of some modern calcareous shells (abs.): Program 1961, Cincinnati meeting, Geol. Soc. America, p. 191–192.

——— 1963, Amino acids in the proteins from aragonite and calcite in the shells of *Mytilus californianus*: Science, v. 139, p. 216–217.

HASS, W. H., and LINDBERG, M. L., 1946, Orientation of the crystal units of conodonts: Jour. Paleont., v. 20, p. 501–504.

HEATH, H., 1911, The Solenogastres: Mus. Comp. Zoology, Harvard Coll. Mem., v. 45, p. 1–182.

——— 1918, Solenogastres from the eastern coast of North America: *ibid.*, pt. 2, p. 183–260.

HERDMAN, W. A., 1910, Tunicata: Cambridge Natural History, v. 7, chap. 2, p. 35–111: London, Macmillan & Co.

HOFFMAN, H., 1929, Aplacophora, *in* BRONN, H. G., Klassen und Ordnungen des Thier-Reichs, v. 3, pt. 1, p. 1–134.

HUDSON, J. D., 1962, Pseudo-pleochroic calcite in recrystallized shell-limestones: Geol. Mag., v. 99, p. 492–500.

KUENEN, PH. H., 1950, Marine Geology: New York, John Wiley & Sons, 568 p.

LABBÉ, A., 1933, Sur la présence de spicules silicieux dans les téguments des oncidiadés: C.R. Adad. Sci., Paris, v. 197, p. 533.

LEMCHE, H., 1957, A new living deep-sea mollusc of the Cambro-Devonian class Monoplacophora: Nature, v. 179, p. 413–416.

——— and WINGSTRAND, K. G., 1959, The anatomy of *Neopilina galatheae* Lemche, 1957 (Mollusca Triblidiacea): Galathea Rpt., v. 3, p. 9–71, Copenhagen.

LOWENSTAM, H. A., 1954*a*, Environmental relations of modification compositions of certain carbonate-secreting marine invertebrates: Natl. Acad. Sci. Proc., v. 40, p. 39–48.

——— 1954*b*, Factors affecting the aragonite-calcite ratios in carbonate-secreting marine organisms: Jour. Geology, v. 62, p. 284–322.

——— 1954*c*, Systematics, paleoecologic and evolutionary aspects of skeletal-building materials, *in* Status of invertebrate paleontology: Mus. Comp. Zoology, Harvard Coll. Bull., v. 112, p. 287–317.

——— 1955, Aragonite needles secreted by algae and some sedimentary implications: Jour. Sed. Petrology, v. 25, p. 270–272.

Lowenstam, H. A., 1960, Paleoecology (Geochemical aspects): McGraw-Hill Encyclopedia of Science and Technology, p. 516–518: New York, McGraw-Hill Book Co.

——— 1961, Mineralogy, O^{18}/O^{16} ratios, and strontium and magnesium contents of recent and fossil brachiopods and their bearing on history of the oceans: Jour. Geology, v. 69, p. 241–260.

——— 1962*a*, Goethite in radular teeth of recent marine gastropods: Science, v. 137, p. 279–280.

——— 1962*b*, Magnetite in denticle capping in Recent Chitons (Polyplacophora): Geol. Soc. America Bull., v. 73, p. 435–438.

——— and Epstein, S., 1957, On the origin of sedimentary aragonite needles of the Great Bahama Bank: Jour. Geology, v. 65, p. 364–375.

Menzies, R. J., and Robinson, D. J., 1961, Recovery of the living fossil mollusk, *Neopilina*, from the slope of the Cedros Trench, Mexico: Science, v. 134, p. 338–339.

Moret, L., 1943, Manuel de paléontologie végétale: Paris, Masson & Cie, 216 p.

Odum, H. T., 1951, Notes on the strontium content of sea water, celestite radiolaria, and strontianite snail shells: Science, v. 114, p. 211–213.

——— 1957, Biogeochemical deposition of strontium: Inst. Marine Sci., v. 4, p. 38–114.

Pilkey, O. H., and Hower, J., 1960, The effect of environment on the concentration of skeletal magnesium and strontium in *Dendraster:* Jour. Geology, v. 68, p. 203–214.

Porbeguin, T., 1954, Contribution à l'étude des carbonates du calcium, précipitation du calcaire par les végétaux, comparison avec la monde animal: Ann. Sci. Natur., Botanique, 11 sér., v. 15, no. 5, p. 29–109.

Reichel, M., 1956, Sur un Trocholine du Valanginien d'Arzier: Ecologae Geol. Helvetiae, v. 48, no. 2, p. 396–408.

Schmidt, W. J., 1959, Bemerkungen zur Schalenstruktur von *Neopilina galatheae:* Galathea Rpt., v. 3, p. 73–77, Copenhagen.

Siever, R., 1957, The silica budget in the sedimentary cycle: Am. Mineralogist, v. 42, p. 821–841.

Smith, C. L., 1940, The Great Bahama Bank. 1. General hydrographic and chemical factors; 2. Calcium carbonate precipitation: Jour. Marine Research, v. 3, p. 1–31.

Stehli, F. G., 1956, Shell mineralogy in Paleozoic invertebrates: Science, v. 123, 1031 p.

Thompson, T. G., and Chow, T. J., 1955, The strontium-calcium atom ratio in carbonate-secreting marine organisms (papers in marine biology and oceanography): Deep Sea Research Suppl., v. 3, p. 20–39.

Thorson, G., 1950, Reproductive and larval ecology of marine bottom invertebrates: Biol. Reviews, v. 25, p. 1–45.

Troelsen, J. C., 1955, On the value of aragonitic tests in the classification of the Rotaliidea: Cushman Found. Foram. Research Contr., v. 6, p. 50–51.

TUDGE, A. P., 1960, A method of analysis of oxygen isotopes in orthophosphate—its use in the measurement of paleotemperatures: Geochim. & Cosmochim. Acta, v. 18, p. 31–93.
TUREKIAN, K. K., and ARMSTRONG, R. L., 1960, Magnesium, strontium, and barium concentrations and calcite-aragonite ratios of some recent molluscan shells: Jour. Marine Research, v. 18, p. 133–151.
TWENHOFEL, W. H., 1932, Treatise on sedimentation: 2d ed., Baltimore, Williams & Wilkins Co., 926 p.
ULRICH, E. O., and BASSLER, R. S., 1931, Cambrian bivalved Crustacea of the order Conchostraca: Proc. U.S. Natl. Museum, v. 78, Art. 4, p. 1–130.
UREY, H. C., LOWENSTAM, H. A., EPSTEIN, S., MCKINNEY, C. R., 1951, Measurements of paleotemperatures and temperatures of the Upper Cretaceous of England, Denmark, and the southeastern United States: Geol. Soc. America Bull., v. 62, p. 399–416.
VAUGHAN, T. W., 1917, Chemical and organic deposits of the sea: Geol. Soc. America Bull., v. 28, p. 933–944.
——— 1924, Present status of studies on the causes of the precipitation of finely divided calcium carbonate: Nat. Research Council, Rept. Comm. Sedimentation, p. 53–58.
VINOGRADOV, A. P., 1953, The elementary chemical composition of marine organisms: Sears Found. Marine Research, no. 2, 647 p.
WANG, H. C., 1950, A revision of the Zoantharia Rugosa in the light of their minute skeletal structures: Royal Soc. [London] Philos. Trans., ser. B, v. 234, p. 175–246.
WELLS, J. W., 1954, Coelenterata, *in* Status of invertebrate paleontology: Mus. Comp. Zoology, Harvard Coll. Bull., v. 112, p. 109–123.